Interwoven

Interwoven

Junipers and the Web of Being

Kristen Rogers-Iversen

With love to our favorite people!

Kristen Rogers-Iversen

University of Utah Press
Salt Lake City

Utah State Historical Society
Salt Lake City

Copublished with the Utah State Historical Society.
Affiliated with the Utah Division of State History,
Utah Department of Heritage & Arts.

The Defiance House Man colophon is a registered trademark
of the University of Utah Press. It is based on a four-foot-tall
Ancient Puebloan pictograph (late PIII) near Glen Canyon, Utah.

Library of Congress Cataloging-in-Publication Data
Names: Rogers-Iversen, Kristen, 1953- author.
Title: Interwoven : junipers and the web of being / Kristen Rogers-Iversen.
Description: Salt Lake City : University of Utah Press, [2017] | Includes
bibliographical references and index. |
Identifiers: LCCN 2017024794 (print) | LCCN 2017025862 (ebook) |
ISBN 9781607815921 () | ISBN 9781607815914 (pbk.)
Subjects: LCSH: Junipers—West (U.S.) | Junipers—Ecology—West (U.S.) |
Human-plant relationships—West (U.S.)
Classification: LCC SD397.J8 (ebook) | LCC SD397.J8 R64 2017 (print) | DDC
634.9/755—dc23
LC record available at https://lccn.loc.gov/2017024794

All photographs, including the cover image, courtesy of Ed Iversen unless otherwise noted.

Printed and bound in the United States of America.

Contents

Illustrations

Color Plates

Plates follow page 86

Acknowledgments

A person can't begin to list a lifetime of mentors and companions whose influence helps shape a book. However, I can acknowledge the richness that uncounted people have brought into my life.

And I can name those who helped me most directly. My parents, Bill and Donna Smart, first took me onto the Colorado Plateau and over the years have shared many a "juniper" saunter (a word that, as John Muir's charming folk etymology would have us believe, comes from the French *sante terre*—"holy earth"). They also read the manuscript and gave astute advice. I will always be thankful to have had for a companion Randy Rogers, who loved the desert country and joyfully nurtured this love in me and our children. Those children, their children, and now my bonus children have been juniper companions along the way, literally and psychologically. I am grateful for all our saunters, and for every time they and many friends asked, "How is the juniper book coming?" One day, while I was lost and discouraged in the midst of the research labyrinth, a little package arrived; inside were "ghost bead" earrings and juniper pictures drawn by young hands. Other loved ones carved gifts of juniper wood, gathered and strung ghost beads, gave juniper-inspired gifts, and painted juniper trees. Rebekah Smith generously created a drawing to illustrate a story in the book. Loved ones who have in countless ways supported my efforts have truly extended lifelines, and I can't thank you enough.

Beyond his constant collegial (and filial) support, Jedediah Rogers gave me the large gift of carefully reading the first draft and making crucially helpful comments. He and Holly Rogers gave expertise and time to create the index. I am also grateful to the peer reviewers who encouraged, criticized, and offered suggestions. At the University of Utah Press, editor John Alley encouraged and gave wise advice, director Glenda Cotter provided support *and* her impressive expertise on birds, Patrick Hadley shepherded the project and provided etymological expertise, and copyeditor Laurel Anderton offered skill and meticulous care in improving the manuscript. She also provided a quote used in chapter 6, for she wrote a poignant entry in the Jardine Juniper's visitor log that I read many years ago, not knowing how serendipity would bring that very writer to collaborate on this project.

Several experts and people involved in juniper issues were willing to patiently educate someone who came out of nowhere, eager but knowing relatively little. I am most indebted to them. Particularly I would like to mention Tiffany Bartz and Neal Clark of the Southern Utah Wilderness Alliance; soil ecologist Jayne Belnap; forester and environmental scientist Mark Brunson; wildlands and range scientist Fee Busby; Alison Jones of Wild Utah; biologist and forester Ronald Lanner; Dorena Martineau of the Paiute Indian Tribe of Utah; Keith Olive and Vicki Tyler of the BLM; forester and silviculturist Doug Page; and Brant and Betty Wall. I am indebted also to numerous dedicated researchers and writers on pinyon-juniper issues. I hope I have represented them all fairly.

And finally, I thank Ed Iversen, whose enthusiasm for this project never lagged, and without whom I would have not have had the space, the time, and, sometimes, the motivation to bring it to fruition. He joyfully created most of the photographs in this book on countless saunters in juniper country. I thank him for his generosity in reading, brainstorming, advising, providing clarity, and offering "cookies"; and for his boundless love, not only for me but for all our kin.

Prologue

Strands

I love trees—who doesn't?—but junipers don't particularly stand out. They don't please crowds like flowering cherries do, or like giant sequoias or those wide-spreading live oaks in the South. Junipers don't offer anything like almonds or oranges. What could there be to learn about such a plain Jane tree? When I told an acquaintance that I was going to study junipers, she snarled, "I hate junipers!" That kind of ended the conversation. She probably meant the landscaping shrub. But I was talking about the tree that grows across millions of acres in the American West. A lot of people probably hardly notice these trees, but they do inspire emotion in others. One person might despise them for sprawling across potential grazing land, or just for being part of a "vast contiguity of waste" in an arid landscape.[1] Others might love junipers as unique and beautiful life forms, fear them as fire hazards, or get stressed when they move into sage grouse habitat. Those are just a few of the attitudes of our day.

I was actually walking through Douglas-firs when junipers vividly popped into my consciousness. They'd apparently been lurking in some obscure corner of my mind—and now, suddenly, without warning, I wanted to learn all about them. Maybe some shadowy part of my mind had woven together all my experiences with junipers into a project that I didn't consciously choose.

So what would you learn if you investigated beneath, around, and through the surface appearance of one particular thing? What would you discover? To find out, I wandered around juniper groves, both literally and literarily. I followed many winding paths, spent a good deal of time rummaging down rabbit holes, and over time found and followed multiple perspectives, stories, and

Utah juniper in Chesler Park, Canyonlands National Park.

meanings. I began to glimpse the place this tree has held in ecosystems and human lives over millennia. It has acquired supernatural traits in myths and fairy tales. It has warmed, sheltered, healed, and even fed the humans who have lived among the trees. It has played a key role in the human and natural history of western North America, historically, botanically, ecologically, culturally, technologically, spiritually, politically, and scientifically.

On a prosaic level, my wanderings widened my understanding of junipers and how humans have interacted with them. On another level, I began to see how this tangled web of stories, connections, meanings, facts, and relationships is tied to everything else. If we can see one obscure element of the natural world with more expansive vision, we can start to sense that *everything* around us has its own hidden stories, and that they have become inextricably connected to our own; as John Muir said, "When we try to pick out anything by itself, we find it hitched to everything else in the Universe."[2] There are many ways to look at something, many ways to be in relationship with something. At the same time, no matter how many facts we may discover, on one level a juniper tree will remain a mystery, as life itself is a mystery, one strand of the unfathomable entwined universe.

*

Within this multidimensional web, the juniper just *is*. It does what it needs to in order to live and propagate in its native environment.

To get a sense of the variety of these environments in the West, visit in your mind's eye just a couple of juniper places. In a wash of salmon-colored sand on the Colorado Plateau stands a hoary tree, silent in the heavy noon heat. It has endured there for hundreds of years, while humans and other animals have passed through. Ragged shreds of bark hang from its bulky trunk. Massive twisting limbs spread a canopy of shade. At the base of the tree, tucked against the roots, lie drifts of "ghost beads"—juniper seeds with a tiny hole in one end. At the same time, on a west-facing slope in the Great Basin, countless junipers grow thick across a hillside. Beneath and between them a few grasses and flowers grow. Leaf litter and old cones cover the ground. A jay flits through the branches. Within this woodland are jackrabbit scat, lichen-spotted rocks, and a bobcat trap—a couple of feathers suspended from a juniper branch, right above a lightly buried steel leghold trap.

No matter where and how it grows, the juniper's life is interconnected with and affects other members of the ecosystem. A tree may help deer survive winter, shelter rodents among its roots and birds within its branches, produce "fruit" that coyotes eat and scatter, take up water that would otherwise flow differently, crowd out other plants, or grow wood and leaves that humans find useful. It may weaken or die through a lightning strike, insects, disease and parasites, fire, old age, or human actions.

The juniper species of the West have been here for a long time, influencing and being influenced.

*

It was when he acquired a pocket manual on the natural history of the Sierra Nevada that the writer John Tallmadge unknowingly began "a lifelong practice of becoming native: learning to weave the stories in the land together with your own story—human, natural, and personal history all bound up together."[3] That's a practice anyone can take up by making the effort to understand stories of the land and its innumerable inhabitants. These stories have been accumulating—like topsoil, as Wendell Berry says—for centuries.[4] In the American West, many layers of juniper stories have become part of the community soil, mixed in and intertwined with countless other stories. The soil is rich, and if we dig a little the stories will start to become part of our own history, or vice versa, creating more connections not only to trees but also to the entire socio-ecological community.

The other part to becoming native, the essential part, is *being* in the land—and living our own stories. I can say without hyperbole that my many encounters with junipers have shaped me and helped me become "native" to the West. I have been brought to an abrupt stop before wild sculpted trees. I have negotiated miles of trees crowded close. I have walked around stumps and the piled skeletons of trees. I've rested in juniper shade and slept on juniper litter. We're both natives, the junipers and I.

*

The deep history of junipers began millions of years ago. The human relationship with junipers of the American West began some twelve thousand years ago and spans the time from the earliest people surviving on the land to the scientists studying it and managers making decisions about it today. Through juniper stories we can trace the arc of human endeavors in the West, how we have used its resources to survive, thrive, or—rarely—get rich; how we have abused those resources; how we have fought over them; how we have imagined and investigated the world around us and our place in it; how we have found spiritual and psychological meaning; and finally, how all these endeavors change the land, have always changed the land and thus changed us.

I have chosen to intertwine the stories in this book, echoing how living things, events, and ideas intertwine in reality. This is fair warning, in case you were expecting a linear path through juniper facts. Nor do I stick to "just the facts" of juniper. We'll wander off sometimes, glimpsing more about the lives that have intersected with juniper, for instance. Why? Because people didn't just appear in juniper country and then disappear. To see a fuller picture of who and what has been in relationship with juniper—and why—is to understand the West, and human endeavor, more fully.

A book can explore only a few strands of the many stories and the voluminous information surrounding junipers, but it's my hope that we can better understand how "every tree, like every other living organism, is at the center of its own four-dimensional spider web. Tug on this strand or that and see what quivers, what falls, what comes in or goes out, what lives or dies, what grows fat—and when," as biologist Ronald Lanner writes.[5] Different strands of the juniper ecosystem have been tugged—and sometimes broken—by many, in many ways, as we'll see.

Let us follow some strands and see where they lead.

Utah juniper rooted on the cliffs of Stansbury Island in the Great Salt Lake.

1

Roots

Decades ago on a steep and rocky cliff on Stansbury Island, overlooking the ethereally flat Great Salt Lake, a juniper seed fell into a crack in the rock. The growing sprout curved its way upward toward the sun, while roots followed crevices and cracked rock to find pockets of soil and moisture. Today the brawny roots muscle through and around the jumbles of stone, its trunk squeezes through a five-inch slit between boulders, and its branches have taken on the wild shape of wind and blizzards.

The juniper genus has evolved tough and sturdy. In the American West, juniper trees stand their ground through drought, cold, heat, and wind, buttressed and supported by robust roots. The Utah juniper (*Juniperus osteosperma*)[1] can drive its taproot downward as far as fifteen feet. Lateral roots can snake out one hundred feet from the stubby tree they nourish. From these mother roots grow networks of rootlets that absorb the scant moisture available. The roots also capture and concentrate whatever nutrients they can gather from austere soils, making it hard for less aggressive plants to compete in a landscape that may get just ten to fifteen inches of moisture a year. In high deserts, where unpredictable summer rains tend to run off or evaporate, roots take full advantage of melting snow and any cool-season rains.

The roots adapt as needed to keep the tree alive. In times of drought, a young tree pushes its taproot down that much more quickly. If the taproot hits bedrock, it turns and finds another way. When the desert soil is shallow, the roots stay shallow. If a lateral root hits a buried boulder, it dives down, around, and up. Or it finds a crack and works its way through. It may follow

A juniper root works its way down a cliff, Cedar Mesa, Utah.

Roots negotiate boulders, Canyonlands National Park.

a crevice downward in pursuit of water. Curiously, the roots of one-seed junipers (*Juniperus monosperma*), and maybe others, pause in their growing during the afternoons. Instead, they put on length during the nights and mornings, growing most rapidly just before midnight.[2]

The doings of roots are in many ways mysterious.

*

Our human roots are mysterious too, and they influence our interrelationships with the natural world. Ancestral, family, and cultural roots ground us in certain ways of thinking and being. Much of the time we are hardly aware of this underlying network and how it nourishes the life we have grown into—how it brings sustenance, and by what convoluted paths, to our outward habits and thoughts. The roots of our attitudes and perceptions hold our human cultures in place. They connect us, help us adapt to the time and conditions in which we find ourselves. At the same time, they may keep us anchored to narrow worldviews. An example: in Burma, men are free to sit and meditate close to certain statues of Buddha, but women must stand or sit off to the side. Men can add a little square of gold leaf to a Buddha statue as part of their devotion;

women cannot do this. In a taxi in Yangon, I asked the driver why this is so. He stopped at a traffic light and looked around with a puzzled expression—why wouldn't this foreigner know something so basic? With his hand demonstrating, he said, "Woman—lower."

We're not buying that kind of outmoded thinking in our own culture—or are we? We may be continually trying to eradicate attitudes like this, but roots run deep, and we "moderns" still have plenty of unconscious or unexamined assumptions about the world. For me, it's an ongoing struggle to even recognize those assumptions—or to wake up and see the ways that I'm locked inside "a narrow room, and tall, / with pretty lamps to quench the gloom / and mottoes on the wall."[3]

So—what about our beliefs and attitudes about the natural world? Each of us may harbor points of view that are quite different from those of others, but most of us have assimilated cultural biases as part of our own. When I think about the roots of my own attitudes toward nature and my place in it, I find a complex of influences: cultural and family norms; my individual disposition and experiences; art, science, advertising, and writings of all kinds; friends and mentors; and of course "nature" itself. Not nature as an abstract concept, but all the organisms, landforms, weather patterns, resources, bodies of water, air, and more that are part of my experience.

Juniper is only one of these organisms. It has influenced not only me but also people across millennia, and it has become part of the "root system" of widely different cultures. Because the *Juniperus* genus is so widespread, its influence has likewise been wide. Members of *Juniperus* take root mainly in the Northern Hemisphere, though one kind of juniper does grow in East Africa. Common juniper, a low-growing shrub or small tree, may have the largest range of any woody plant, growing from the Arctic Circle down to about 20° north latitude all across North America, Europe, and Asia. Indeed, on a steep rocky climb on Penobscot Mountain in Maine, my trusty and exuberant hiking buddy, Eddie, remarked that this looked like a pretty tough environment—just the place where you might find junipers. We opened our eyes then, and sure enough, almost right at our feet, prickly mats of common juniper were growing among patches of wild blueberries. Surely not many plants grow, as the juniper genus does, from sea level to sixteen thousand feet up in the Himalayas, in such far-flung places

A Utah juniper, surviving where it sprouted on the Velvet Cliffs in Wayne County, Utah.

as Tibet, Greece, Mexico, Kenya, Finland, Haiti, China, Estonia, Greenland, and the Arabian Peninsula.[4] In all these places, juniper has a place in lives and cultures. It has become part of people's roots.

Junipers have ecological roots that are not only wide but also deep. Like us, they have evolved for millions of years. Some fifty million years ago, the junipers became distinct from their cypress relatives, which include redwoods, sequoias, and true cedars. You can still see family resemblances among many juniper species and cedars in their shreddy bark and scalelike needles. There's a reason why in the arid West, where no true cedar can grow, people call junipers "cedars."[5]

*

Several species of juniper are rooted widely in the American West—in its own deep history, ecology, human history, and development. Pinyon-juniper (aka PJ) woodlands cover forty-seven million acres in the five states of the U.S. Southwest, making up 53 percent of all forest cover in these states. Below the pinyon-juniper belt grow additional pure juniper woodlands.[6] Juniper thrives in the West because it doesn't mind rough neighborhoods like cliffs, talus slopes, mesas, arid hills and canyons, and alkaline soils between 4,500 and 6,000 feet in elevation, where the sun beats down in the summers and freezing winds blow in the winters. In his classic *Natural History of Western Trees*, Donald Peattie describes the Utah juniper and its harsh surroundings. In so doing, he also mythologizes and stereotypes the tree, showing how we instinctively link nature with our own well-rooted human constructs: "In its namesake state of Utah, [Utah juniper] is as characteristic a settler as the Mormons, and in its venerable age sometimes reminds you of an old patriarch of the sect—rugged and weathered and twisted by hardship, but hard too to discourage or kill. . . . No other tree is so well-fitted to endure the arid, wind-blown, sand-swept land of Deseret."[7]

The juniper has adapted so well to the desert—in its leathery, gnarled way—that a person used to a more succulent landscape could hardly think it of much use. But in countless ways the juniper has played its own unique role in the West. It has aided humans in numerous ways and has vexed them too—but that depended on what they were trying to do. If you were indigenous, you would know many ways to use juniper to make life easier. If you were near dying of heat and thirst, you would look upon juniper shade as a true blessing. If you felt connected to invisible and ineffable qualities in nature, you might seek spiritual aid through the tree. If you were a rancher you could appreciate the rot resistance of its wood for fences but hate the fact that cows can't find much to eat in a juniper woodland. If you were a land manager trying to do right by the land and various stakeholders, you might weed out these woodlands in favor of grasses. If you were a botanist, you might see in the tree a complex and fascinating subject.

The range of reactions to juniper is plain to see in this essay excerpt about western juniper (*Juniperus occidentalis*), a species that grows mostly in Oregon and California:

> It's a plant with a dual identity. At times it plays the wizened monarch of the rimrock, its girth and gnarly visage testaments to its age. At other times, and in other places, it's the upstart, the bully. It moves onto land it did not

> previously occupy, ousting the long-standing inhabitants, and wreaking havoc in an otherwise nice neighborhood.
>
> Humans venerate the old beaten up snags and barely living remains, which inspire them to designate wilderness areas (e.g., Oregon Badlands near Bend), create photographic art, and write poetry; in contrast, expanding populations of younger trees are the target of wholesale slaughter. It seems that people either love this juniper or hate it, really hate it.[8]

Some people consider the PJ (pinyon-juniper woodlands) "public enemy number one." As one landowner in Texas said, "Cedar [meaning juniper] deserves the reputation as the vilest plant thriving in Texas."[9] Others, as one land manager noted, consider these woodlands "our very life! They are part of us."[10] Edward Abbey said, "If my decomposing carcass helps nourish the roots of a juniper tree or the wings of a vulture, that is immortality enough for me."[11]

Clearly, these opposite attitudes are rooted in different soils altogether.

*

A few years ago I talked with Ken Yamane, a bonsai master who sometimes dug up Utah junipers (with permission from the Bureau of Land Management) to shape into bonsai. "The trick is to find a tree one to two feet tall and that is one hundred years old," he told me. "They're really hard to find. A deer might have eaten it, or a truck came by and ran over it and broke it off." When he dug a tree, Yamane would cut off the long taproot and then trim the long horizontal roots as well. It would take an hour or two to dig up one small tree. "You have to be very careful. It's just like us—the older we get, we're not as healthy as a twenty-year-old. So with the older trees you have to be kind of careful."

During the first year, you just wait to see if the tree will survive, which is rare; another bonsai enthusiast told me he had dug up fifty Utah junipers and only two of them had lived. But if it does survive, you begin to train it. After a few years, when it looks the way you want it to, you pot it. A few years later, you might need to trim the roots, or put the tree in a bigger pot. Yamane said that training and caring for a tree for so many years creates a bond. "I guess they're like my dogs, or almost like my kids. You spend so much time; they're kind of special to me."

Yamane's relationship with juniper is of course rooted in his own Japanese American culture and family; his father first showed him how to grow bonsai. During World War II, U.S. government officials had come to the elder Yamane and asked if he would be willing to enlist. Yes, he said, as long as he wouldn't have to fight his brothers. "I guess that was the wrong answer, so he got sent to Tule Lake in California"—one of the ten hastily built internment camps where the government imprisoned Americans whose only crime was to be of Japanese ancestry. Ken Yamane was a baby at Tule Lake. After the war, his family moved back to Japan for a while. There, he absorbed a passion for his culture, particularly the patient art of bonsai. While today people usually use wire to shape trees, Yamane watched his father use chopsticks and rocks hanging from branches to nudge the small trees into harmonious forms.

When they returned to the United States in the 1950s, the family again encountered racism and hatred rooted in their small Utah community. "All through high school, some kids wouldn't talk to us. One of the neighbors said, 'Don't come around our house,' because she said we're possessed by the devil." With bonsai, Yamane was living his desire to help Japanese Americans embrace and preserve their own roots. His dream, he said, was that every Japanese family in America would have a bonsai tree in the back of the house.[12]

*

If during the 1930s Utah politicians had known or considered how much junipers are part of westerners' roots, they might have considered making the Utah juniper the state tree. And then again, maybe they wouldn't have. A state tree needs to project a certain image. It needs to communicate something about the character of a state and its people. And when the Great Depression has sucker punched your state, you need an image that will inspire respect. Juniper wasn't on the radar screen—although, actually, the box elder was. A legislator from Box Elder County proposed that tree as the perfect symbol of Utah. Aroused, the Poplar Grove neighborhood in west Salt Lake City parried by suggesting the poplar. Those particular promotional efforts failed. After fending off several bids by box elder fans, the stately blue spruce (aka the *Colorado* blue spruce) won the title in 1933. Utah now had a symbol that, supposedly, could communicate the state's strength, beauty, aspirations, and nobility of character. (The state of Colorado uses *its* state tree, which is also the Colorado blue spruce, to project these same attributes. But to be fair, Utah got it first. Colorado didn't adopt its state tree until 1939.)[13]

Decades later, some people had the nerve to suggest that the Utah juniper would symbolize the state better than the spruce. There's the name, of course. And then there's the fact that no other tree has as many individuals or covers more territory in the state than the Utah juniper. Besides, some people rather like the craggy, humble image of the juniper. It's a fine image—but not the kind of image Utah boosters have generally wanted to project.

In 2008 some fourth-grade teachers rallied students to "Jump for the Juniper" to become Utah's new state tree. In their Utah Studies classes, these kids learned about the tree, and they learned how to lobby. They posted drawings on a website, with arguments along these lines:

> "Right now our state tree is the Colorado blue spruce. But I like the Utah juniper. It has been helpful to indians [*sic*] and other early people."
>
> "The Utah juniper is a lovley [*sic*] tree and the Indians used it for most of their needs. So that is why I want to change our state tree!"
>
> "The blue spruce is pretty but so is the juniper. The Utah juniper is as pretty as the people of Utah."[14]

The fourth graders and teachers showed up at a hearing of the state legislature's committee on natural resources, agriculture, and environment to support a juniper-as-state-tree bill. Fourth-grader Sally Devitry spoke before the lawmakers, and the students showed a video they had made. The Farm Bureau backed the kids up. But the Utah Cattlemen's Association couldn't do that. Years of fighting environmentalists over projects meant to expand rangelands by removing junipers had made the members of the association nervous. Give the Utah juniper special status, they apparently believed, and environmental groups would have more ammunition to stop juniper removal projects. Even though the bill explicitly stated that the juniper would receive no special protections, the ranchers were probably right to feel threatened by the power of perceptions versus facts. So even though it meant going toe-to-toe with little kids, the Cattlemen's Association spoke out against the bill. Over these objections, the subcommittee went along with the students and voted unanimously to recommend the bill.[15] But this was politics. After the kids went home, the full legislature voted with the cattle industry, and the spruce kept its job.

Whether or not a state tree makes much of a difference in how outsiders view Utah, some people put a lot of stock in state symbols. All states draw from nature to define themselves. Utah's symbols include the sego lily, Indian ricegrass, elk, and of course the California gull. Sagebrush, bristlecone pine, mountain bluebird, and desert bighorn sheep represent Nevada. Arizona has the saguaro blossom, ringtail, cactus wren, and Arizona tree frog. New Mexico has the yucca flower, pinyon pine, roadrunner, and black bear. Wyoming: bison, meadowlark, Indian paintbrush, plains cottonwood. Colorado: lark bunting, columbine, Rocky Mountain bighorn sheep. Idaho: mountain bluebird, syringa, huckleberry, western white pine. Montana: bitterroot, grizzly, ponderosa pine, meadowlark. Even these short lists give a sense of different western ecosystems and the ways people use the natural world to express who they are. Our cultures, our lives, and in many ways our attitudes are rooted in our ecosystems.

In 2014, schoolchildren again lobbied the Utah legislature to change the state tree—this time proposing quaking aspen. In the natural world there is no hierarchy of importance. Very different trees like spruce, aspen, and juniper each have roles to play in their ecosystems. But a state "needs" a tree symbol. This time the Utah legislature got on board with the aspen, a tree that is actually declining throughout the West and poses no threat to special interests. The children argued that aspens, which grow as colonies of trees with a single root system, represent how Utahns "work together to reach new heights." The bill's sponsor, Senator Ralph Okerlund, liked how "the quaking aspen provides Utah an economical, agricultural and recreational benefit." The strong root system also reminded him of "the state's [or, in other words, Mormons'] emphasis on family history and genealogy." Representative Brad Wilson noted, "The aspen stems grow from roots of older trees. This creates a very important metaphor that we could connect to Utah as these children of the parents grow and are very prolific."[16]

The fact that the Colorado spruce makes up just 1 percent of Utah's forest cover and the aspen makes up 10 percent and grows in every county apparently also influenced the legislature's decision to replace the spruce. Of course, if percentage of cover and distribution were truly the deciding factors, Utah juniper would win hands down. But the juniper certainly can't provide the same kind of communal roots metaphor that aspen can. For instance, the 106-acre grove of quaking aspen in central Utah named Pando has a root system that may be eighty thousand years old. This single organism (the interconnected roots along with thousands of stems) is thought to weigh, altogether, more than thirteen million pounds—perhaps the heaviest organism on earth.[17] Those are mighty roots indeed.

Juniper roots, on the other hand, might represent individuality, resilience in hardship, and resourcefulness. When they chose aspen, the legislators elevated values like community, shared endeavors, genealogy, economic development, agriculture, and large families. Over time, this choice could subtly affect the way Utahns think about themselves and nature. Not in a big way, of course. Not even in a perceptible way. But the meanings we give to nature will always be a part of our cultural roots, and those roots will always sprout into something.

2

Germination

One year, my mother and I had the opportunity to travel to Many Farms, Arizona, with a small band of volunteers for Adopt-a-Native-Elder, an organization that brings food and supplies to Navajo elders twice a year. The elders look forward to this. Their children and grandchildren bring them to a central location for a lunch of fry bread and beans. They all receive a bag of Bluebird flour and boxes filled with things like oatmeal, soup, Mentholatum, and coffee. On this warm autumn day, the Diné women were bright with velvets, satins, and turquoise jewelry; the men in jeans, cowboy shirts, and caps or hats. The volunteers served food and cleaned up, stood through a long ceremony mostly in Navajo, and then carried boxes to trucks. Clinton and Sally Clah had come to this gathering. These two aged people lived in a solitary cinder-block house and nearby hogan, surrounded by miles of desert underneath a wide-open sky. You could reach their place only by successfully navigating a web of rugged jeep trails. We had visited them there once, in their home among the junipers.

On that day at Many Farms, as the afternoon gathering came to a close, Sally Clah shuffled up to me. Such a small woman: the top of her head reached my clavicle. She wore her white hair in a traditional bun, and her dark eyes were still sharp and quick. She could speak no English, but she had never had much to say to us in Navajo even when her nephew translated, so we had never really talked. In her stillness I had felt a little clumsy and intrusive. She didn't speak now. She held out her frail hand and in it was a necklace made from juniper seeds, strung on a cotton string.

*

The inconspicuous female cone of a Utah juniper.

It may take more than thirty years of growing before a Utah juniper begins to produce seeds. How it makes seeds is straightforward conifer sex—mostly. Utah junipers are usually *monoecious*—which means that both male and female cones grow on the same tree. Other junipers in the West are *dioecious*, with separate male and female trees. Some 10 to 15 percent of Utah juniper trees are also dioecious, but that's not the end of the story. We know more about this story because some people actually study the sex life of trees, and Utah juniper is one of the trees people have studied. So what the researchers have noticed is this: some years, a tree that usually has all female or all male cones might switch to having both. Two trees in the study changed sexes altogether. Nobody knows the secret life of junipers for sure, but one hypothesis is that a tree may switch sex when stressed by drought or some other hardship.[1]

The tiny male cones emerge on the ends of the twigs. In the spring they release fine pollen into the wind. The tender green female cones have a few pointed scales that, when fertilized, gradually grow together into a small fleshy structure one-third of an inch long. These cones look like berries, and that's what most people call them. A blue-white blush covers them, and the tips of the scales still poke out. The thin layer of flesh surrounding the seed is full of volatile oils. The flesh smells and tastes resinous, a little like turpentine.[2]

During the first summer, the cones reach full size. But inside, the seed needs another summer to fully mature, and the cones stay on the tree. The seed develops until it is a quarter inch long and stone hard (or bone hard—*Juniperus osteosperma* means "bone-seed juniper"). The mature "berry" is red brown with a pale blue blush, and the seed inside is two shades of shiny brown. Its sides angle to a point.[3]

*

When Sally Clah gathered the beads she held out to me, she would have searched beneath shaggy trees to find each hard little juniper seed with a tiny hole nibbled in one end by a rodent or insect getting at the meat inside. With difficulty, she would have worked a hole into the other end and then strung the seeds, one by one. I had heard that Navajos put necklaces of juniper seeds around the necks of their children to protect them. As she stretched her arms up, I realized, with a sudden rush into my heart, that Sally meant to give the necklace to me, like those grandmothers who for generations had placed beads on their grandchildren. I bent; she put the juniper beads over my head. I didn't even know how to say thank you in Navajo, so how could I tell her what that blessing meant to me—or would mean over the coming years?

Navajos have valued juniper bead necklaces as long as anybody knows or remembers. They hold a sacred place in the Navajo worldview. Folklorist Barre Toelken, who lived with a Navajo family for two years during the uranium rush of the 1950s, learned about this firsthand. In that community, small children would look for stashes of seeds beneath junipers and gather only the ones with holes in them, making sure to leave the whole seeds behind so the animals would have something to eat. No one in Toelken's Navajo family went anywhere without carrying some beads, he said, because they would prevent nightmares and keep a person from getting lost. His Navajo sister told him that this is so because the beads "represent the partnership between the tree that gives its berries, the animals which gather them, and the humans who pick them up. . . . It's a three-way partnership—plant, animal, and man. Thus, if you keep these beads on you and think about them, your mind, in its balance with nature, will lead

"Ghost beads," believed to provide protection to the wearer.

a healthy existence." That makes perfect sense to me personally and societally: if we are in a balanced relationship with nature (which includes ourselves and other humans), we should have good dreams and health, and we won't get lost.[4]

In 1968, Richard Movitz, a ski racer who had competed in the Winter Olympics in the 1940s and later became a jewelry distributor, came across some Navajo women in southern Utah selling the beads. As he fingered the "ghost beads," as white people call them, a marketing idea formed in his head. These juniper beads, he realized, were unique and virtually unknown. But they could really appeal to the hippies of the late 1960s, coming as they did directly from the earth and Native American culture. Ghost beads could make great "love beads." In fact, Movitz dreamed that they might well take off in the general fashion industry of the day, which—"like hippies—grooved on the American Indian look." Movitz got a plan together. He talked to government officials in Utah about how marketing ghost beads could stimulate economic development among the Navajos and cut down on welfare. The state agreed to subsidize a pilot project, and Movitz hired Navajo women to make necklaces that he sold for three dollars each. Some months later, he reported that he had thousands of orders and was having a hard time keeping up with the demand.[5]

*

The scales of juniper cones (commonly called "berries") remain fleshy and fused, covering the seed. Mature Utah juniper berries are covered with a pale blue bloom.

Did these love beads bring about more love, peace and flowers, and financial security? Maybe. I'd be the last person to discount the power of the "juniper's eyes," as Navajos call them. But the power of their seedhood is just as magical and awe inspiring as any esoteric power. Out of this stony little kernel comes a living thing, which, given enough time, sun, water, wind, and seasons, will grow into a uniquely magnificent tree. The strong, twisted tree lies latent in that tiny seed. In that latency the processes of growth proceed slowly. After Utah juniper seeds mature for two summers, they lie dormant inside their tough coats for yet another eighteen months before they can awaken and sprout.[6]

The fleshy cones we call berries enclose those dormant seeds, and they stay on the branches through winter, waiting for a ride to somewhere else. During the stark winter months, the bluish berries are advertising, "Eat me." And birds do. Animals like jackrabbits, coyotes, foxes, pronghorns, and bighorn sheep do. As they meet their own needs for sustenance, all these creatures fill a niche in the workings of the ecosystem: they liberate seeds from the fleshy cone—the intestinal journey will help them germinate more reliably and quickly—and drop them far and wide, spreading the juniper children beyond the parent tree. If you see rows of small junipers right beneath a fence, or if you see a juniper sprouting beneath a ponderosa, you can guess that juniper-loving birds once roosted there.[7]

A juniper seed germinates best when covered with a little soil and when it is gently moist for an extended time, so that the stony seed coat can soften. The processes of evolution have provided the juniper seed with a growth inhibitor that prevents it from sprouting in dry soil, thus avoiding an untimely death. Moisture must be present to leach away this growth inhibitor, and then the seed can take off. Without moist, soil-blessed conditions, a seed can sleep for some time. A 1959 study found that 17 percent of Utah juniper seeds could still sprout after forty-five years. All that time the potentiality of *tree* was latent and invisible.[8]

*

If there is a collective unconscious, human relationships with plants must surely live there, latent. Millennia of interactions with the plants of the earth fold into our collective hearts and lie there, dormant but alive. In the case of juniper, which flourishes so widely and so persistently across the globe, seeds of our ancestors' past may lie beneath all the infinite details of our own slice of time. Who knows? Maybe this is why Edward Abbey felt a recognition, a connection to a particular juniper at Arches National Monument:

> My favorite juniper stands before me glittering shaggily in the sunrise, ragged roots clutching at the rock on which it feeds, rough dark boughs bedecked with a rash, with a shower of turquoise-colored berries. A female, this ancient grandmother of a tree may be three hundred years old. . . . I've had this tree under surveillance ever since my arrival at Arches, hoping to learn something from it, to discover the significance in its form, to make a connection through its life with whatever falls beyond. Have failed. The essence of juniper continues to elude me unless, as I presently suspect, its surface is also the essence. Two living things on the same earth, respiring in a common medium, we contact one another but without direct communication. Intuition, sympathy, empathy, all fail to guide me into the heart of this being—if it has a heart.[9]

Ginevra de' Benci, painted by Leonardo da Vinci around 1474, with a *Juniperus communis* bush behind her. Courtesy National Gallery of Art, Open Access Policy.

Abbey isn't the first or the last to try to plumb the essence of the juniper, on this or other continents. For whatever reasons, this genus of tree resonates in the human psyche. Old stories reveal something of humankind's relationship with the juniper and also with the natural world. The stories show what was important to the people who told them and how they interpreted their lives. The seeds created by those relationships are still germinating within our cultural ecosystem.

For instance, Abbey's grandmother tree is an echo of the sacred mother tree revered in many cultures. In the Middle East, trees were associated with Asherah, a term that could refer either to the great mother goddess of the Semitic peoples or to her representational cultic object (a tree or a pole representing a tree). And in fact, some scholars associate juniper with Asherah.[10] Recent scholarship suggests that the Israelites themselves at one time revered this goddess as the consort of Yahweh.[11]

Tradition (aided by various Bible translations) has it that when Queen Jezebel promised to kill the prophet Elijah, he ran into the desert and rested under a juniper tree. The original Hebrew word used in the story actually refers to the broom tree (*Retama raetam*), but nevertheless we have it in our consciousness that it was beneath a juniper that an angel found, comforted, and fed Elijah. Mother tree.

Again, legend has it that a juniper saved the lives of Joseph, Mary, and Jesus as they fled to Egypt. As soldiers approached the family, the "Madonna's Juniper Bush, as they call it in Sicily, was growing near the wayside, and recognizing the danger of the Lord of Heaven and Earth, opened its thick branches and enclosed within its sheltering arbourage the Holy Family. . . . Its gracious service was rewarded, they tell, by the precious virtues it then engendered for human ills, rendering it henceforth beloved by men and welcomed in the druggist's store."[12] Mother tree, tree of life.

Thereafter, juniper consciousness became intertwined with the Virgin Mary. At Christmastime, Europeans would hang juniper boughs in their stables and near their crucifixes.[13] In 1474 a young Leonardo da Vinci painted the fetching but morose-looking sixteen-year-old Ginevra de' Benci, whose name, Ginevra, is a cognate of the Italian word for juniper, *ginepro.* Behind her stands a common juniper tree, which has prickly leaves instead of the scalelike leaves found on trees in the western United States. On the reverse of the portrait, sprigs of laurel and palm wreathe a sprig of juniper, all tied together with a scroll inscribed *Virtutem Forma Decorat*, meaning "Beauty adorns virtue." It seems that the juniper here, as Mary's plant and as a symbol of the young lady Ginevra, stands for chastity and virtue.[14]

*

For most people, the seeds of juniper consciousness lie quietly. Sometimes they sprout into stories that echo with generations of juniper-human relationships, or, these days, into esoteric claims. For instance, I have found these assertions:

- Ancient Germanic people revered junipers and regarded them as portals to the invisible realms of spirits, fairies, or giants. People would leave offerings to the local nature spirits beneath the trees.[15]
- Hopi Indians believed the earth's "caretaker spirit" travels with the juniper.[16]

- Several cultures believed juniper smoke could enhance clairvoyance and facilitate contact with spirit worlds.[17]
- Norwegians burned juniper branches during childbirth, counting on the sparks to protect the mother and baby from witches and *huldrefolk*, or "hidden people."[18]
- Cutting a juniper tree in Wales was a dire thing to do, since the Welsh believed the cutter would die within a year.
- In the language of flowers, juniper symbolized "perfect loveliness [figure that one out!], succour, and protection."[19]

Junipers have symbolic and mythic meaning in the American West as well. In a story of the Yavapai people of Arizona, a young man sits beneath a juniper skinning a deer. Suddenly a naked young woman jumps from the tree and demands intercourse. Not every young man would run away in such an event, but he does. He escapes and hides under the ashes—juniper ashes, no doubt—of his grandmother's fire. As the story continues, he gets rid of the sharp teeth that she happens to have in her vagina and then, feeling somewhat safer, marries her. However, this same seductive wife later sends her husband hunting in bear country; she herself is a bear and wants her brothers to kill him. But he's no dummy. He makes a decoy by dressing a juniper stump in his clothes. With the aid of this decoy, he kills the bears.

As anthropologist Patrick Morris points out, the Yavapai have used the juniper extensively for building, cooking, heating, sleeping, and more. The tree has been integral to domestic life. So what does this story mean? The dangerous wife emerging from the juniper, Morris thinks, may suggest some ambiguity toward or fear of marriage and domestic life.[20] But a tree that is so vital to life must also have connotations of security and comfort. Only those people who truly depended on junipers in multiple ways—the people who actually told the story—could understand its layers of meaning.

*

I don't think any story shows the power of juniper in cultural undercurrents better than "The Juniper Tree," a folktale recorded by the Grimm brothers. Imagine a powerful story being told around a fire, under brilliant starry skies in the distant past. The story's evolution over space and time may have obscured its original shape and meaning by the time the Grimm brothers wrote it down as they found it in the nineteenth century. But in this mysterious tale, juniper still reverberates with the same qualities and themes it has taken on in other traditions: magic, shelter, gateway to other realms, sexuality, a deadly woman, domesticity, and motherhood. It is a dreadful story.

The Grimms called the tale "Von dem Wacholder" ("From the Juniper"), implying that the juniper is the source of something. And it is. *Wacholder* means "juniper" in German, but it once also carried the meaning of "watchful tree," or "one who watches over."[21] The tale is set some two thousand years ago. It goes like this: A rich man and his beautiful, pious wife long for a child. Her prayers go unanswered until one winter day, while sitting beneath the juniper tree in their courtyard, she peels an apple and cuts her finger. As the blood drips, she wishes again for a child. Then she feels calm and certain it will happen—as if the juniper tree has given her special knowledge. She becomes pregnant.

The months pass. During the fifth month, as she stands beneath the juniper her heart jumps for joy at its fragrance, and she falls on her knees. The next month the juniper berries grow plump. In the seventh month, she picks berries and eagerly eats them. In the eighth month, she cries to her husband, "If I die, bury me beneath the juniper tree!" We might ask what is happening between the woman and the tree to create such an powerful bond. What was it about a juniper, specifically? The narrators provide no answer; the story only gestures toward what the juniper meant to those who told it over the generations.

When a boy is born, the woman does die—of intense joy. A jealous and evil woman then marries the widowed father. When the boy and his stepsister are older, the stepmother schemes about how to kill him. She sends him to get an apple from a trunk, and as he reaches inside for one she slams the lid down, severing his head. The mother then puts the head back on the body, sets it on a chair, and tricks her daughter into slapping the boy, which knocks his head off. The little girl is horrified to think she has killed her brother. In the best ghoulish folktale tradition, the mother cooks her stepson into a stew and serves it to his father.

Sobbing, his little sister gathers up the bones in her best silk scarf and lays them beneath the juniper. She herself lies down beside them. Here beneath the

In a Grimm Brothers folktale, the soul of a murdered boy rises from a juniper tree. Drawing by Rebekah Smith.

branches, the tree's mysterious power washes over her. She feels calmed and comforted, relieved of her terrible guilt and grief. The juniper then begins to undulate, branches moving apart and then together in waves, "like hands clapping"—or perhaps like a mother's body in labor. A mist with a flaming core rises from the tree, and out of the mist flies a beautiful bird, singing gloriously. The tree—the harbor of the pious mother in life and death—has given birth to the child's transfigured soul.

The bird sings a beautiful song among the villagers (actually, the words are horrific; they are about his own murder). The people, enchanted, pay him for his song; he receives a gold chain, red shoes, and a millstone. He returns to the juniper tree. His family is inside, eating dinner (eating is a big part of this story!), so he sings his beautiful song. When the father comes out to listen, the bird throws the gold chain down to him. When his sister comes out, he throws down the red shoes. But when the stepmother comes out to receive her gift, he throws down the millstone, crushing her.

Then, out of smoke and flames the boy emerges, fully reconstituted. The mysterious mother tree has rebirthed the child. He, his father, and his sister all go back into the house and finish their dinner.

In this tale, the protecting, sheltering tree is also the portal to the spirit world. It fulfills desires; it gives life and death; it facilitates justice and restores what has been taken. And in fact, German legend has it that the juniper actually has a helping spirit, Frau Wacholder, or the Watchful Woman; tradition said that you must take off your hat in respect and greeting when passing a juniper.[22] Legends say the tree can restore to you things that have been stolen from you, or that it is a wishing tree. "The machandelbom (juniper) in our fairy-tale is a wishing-tree," writes Jacob Grimm, and "so is that from which Cinderella shakes down all her splendid dresses; the [East] Indians call it *kalpa vriksha* (tree of wishes) or *Manoratha-dayaka* (wish-giving)."[23] According to the Tajikistan Travel Guide, near Iskanderkul Lake is "an old Archa or juniper tree with thousand of ribbons, pieces of fabrics and even cloth labels tighten over. Visitors tight the ribbon and make wish which will of course come true another day."[24]

*

These stories can only scratch the surface of the mystical and mythological powers that humans have vested in the juniper over the millennia. But they give an idea of the relationships humans have created with the natural world, and of the stories that help us work out meanings and values. Mountains, rocks, flowers, birds, lakes, rivers, fish, animals, and trees all have layers of stories, mostly buried now in our collective past and largely unknown and unconscious. Through a few ancient stories that have survived in some form we have dim portals into other times and ancestral people, when the world was more magical, more alive than it is for many of us today.

Still, for each of us, there may be particular trees or landscapes that we invest with power and meaning. Our personal association with the natural world animates the stories we tell about it. It might be an apple tree that a grandfather planted. Or it might be the song of a canyon wren, whose cascading notes during many desert dawns now evoke layers of experiences. It might be a lake, mountain, or beach holding the stories of generations of a family who have gathered there. Such places and living things come to symbolize community, tradition, and shared stories. The Mail Tree, a gigantic cottonwood in Fruita, Utah, comes to mind as one example. This tree, where the men carrying mail by horse and wagon left the community's letters, still stands in Capitol Reef National Park, its community meaning still remembered by those who never lived in Fruita.

Maybe we carry entire places in our bones, places where all the plants, animals, waters, and landforms are woven into the stories of generations.

*

And so now, coming from Tajikistan and the bright strips of cloth fluttering on an old tree, coming back from many landscapes and relationships, back to the West, where Edward Abbey had *his* particular juniper and where Sally Clah strung her beads, we stop at Sixth East in Salt Lake City. There, a small peristyle monument with white columns stands on the grassy median that runs between Third and Fourth South. This shrine, erected by the Daughters of Utah Pioneers (DUP) in 1933, once sheltered the remains of a revered "Lone Cedar Tree." The name comes from the memoirs of John R. Young, a nephew of Brigham Young who came to the Salt Lake Valley as a boy. He described a particular juniper tree: "From our cabin in the mouth of City Creek canyon, in 1847, one could see a lone cedar tree on the plain southeast of us, and on the south fork of the creek [City Creek], about where Main and Third South Streets intersect, stood seven wind swept, scraggy cottonwood trees. On the north side of City Creek stood a large oak tree. No other trees were visible in the valley."[25]

The lone cedar tree idea grew in evocative power. It grew within the collective consciousness of Utah Mormons; children heard about it from their elders, and families made informal pilgrimages to visit the site where this tree grew. Legends evolved over the years. In 1876, an article about a streetcar line extension simply mentions the "old Cedar post corner" on Sixth East.[26] But by the 1920s, the newspapers had quite a bit more to say about that old post. The *Deseret News* noted that the 1923 Pioneer Day celebration would include an outing to "Cedar Post, a relic mark [that] designates the place where once a large cedar tree grew. At one time this tree was the only one of its kind in the valley it is claimed, and acted as a guide post to the old camp ground."[27]

The next year, the *Salt Lake Telegram* reported that the DUP planned to erect a fence around the post to protect it—and then waxed eloquent: "How long since [the] old cedar post was a sapling and how long it stood there before first looked upon by white man cannot be determined, but that some of the gravest secrets connected with the building of the great empire of Zion is locked up somewhere in its heart is not such a matter of conjecture.... It is not too much to presume that many important meetings between outstanding characters took place in its shade, to say nothing of the whispered words of love that must have been exchanged at this naturally popular trysting place."[28]

In 1933, the DUP went a step further, protecting the "last remnant of the cedar tree" beneath the shelter of a white shrine designed by Young and Hansen architects—Lorenzo Young being a grandson of Brigham Young. The peristyle still stands on an island in the middle of Sixth East, as does the "The Cedar Tree Shrine" plaque put up by the DUP at that time. It says that "pioneers entering the valley stopped here to sing songs and offer prayers of gratitude. Later, loggers going into the canyons, trysting lovers, and playful children all enjoyed the shade and protection of this cedar tree." Here's the Mother tree again. "Because of its friendly influence on the lives of these early men and women we dedicate this site to their memory." At the dedication in 1933, five hundred attendees heard words of praise and then sang "Come, Come, Ye Saints," the classic Mormon hymn of faith, resignation, and gratitude.[29]

And so the remains of the "tree" stood honored for many years—it's a stretch, but one might think of the revered Asherah poles—until the night of September 21, 1958, when vandals cut down the trunk and hauled it off, leaving a stump a couple of feet tall. This was big news in 1958, upsetting to many, but not to all. A reporter for the *Deseret News* sought out Utah State Historical Society director A. R. Mortensen for an interview. Mortensen didn't mince words. Since the tree was a "historical fraud," he said, he was "not shedding any tears over its loss. . . . It's only an old dead stump with little historical value." He explained that it couldn't have been the only tree in the valley, as the "lone cedar tree" legend implied.[30]

Mortensen, though eager to set the factual record straight, did not respect the power of the tree in the community memory—those seeds of history and myth that had germinated into deeply vested meaning for Mormons in particular. An editorial in the September 25 *Deseret News* reaffirmed that power.

> Hallowed tree or old dead stump, it symbolized a lofty idealism, a monument as it were, to a tremendous feat, the greatest mass exodus in modern times and the rest and the relief that came to these weary travelers. As long as so many believed in the tree, it was an important item in the history of their most valiant forebears. There is no legend, no tradition without belief. . . . It was their friend, whether it was the original tree or another like it. It matters little. The Lone Cedar Tree represented kindness, shelter, hospitality—all given freely and withheld from none, redmen or white.[31]

Mortensen paid for his candor in an ideological battle with Kate Carter, the formidable president of the DUP, who "publicly took me to the woodshed," he later wrote. Carter was an enthusiastic defender of the ideal of noble Mormon pioneers and a collector of pioneer "relics," a far different personality from the professionally trained Mortensen—and the sparks flew between them. Carter didn't succeed directly in getting Mortensen fired, but he left the Historical Society soon after. In the meantime, the DUP put up another plaque that reminded all, "In the glory of my prime I was the pioneer's friend." In the end, eulogizing Carter at her death, Mortensen said she was "a great and noble lady who, although she could not apologize for 'skinning me alive' invited me to be the principal speaker at her annual luncheon the very next year. . . . We were close friends forever after."[32]

We may laugh about that little tempest now—especially in the ironic light of an article in the June 1, 1919, *Salt Lake Tribune.*[33] The writer reported on the pending demolition of a house at 576 East Third South and said that the "Old Cedar Post" in front of the house was to be removed. According to the legend perpetuated by this article, Brigham Young had stuck the post into the ground to guide pioneers entering the valley. The woman who lived in the house in 1919 had "taken care of the post" for more than thirty years. She had heard various stories about its origin, including the one about it being the remnant of the only tree in the valley. But after a storm broke the wood, she dug around the base to see whether any roots remained. Instead of roots, she found a sawed-off end. Sure enough—it *was* a post, not a tree stump. Perhaps it had once been *the* tree and had been sawed off and moved. But who knows now?

The woman recommended to the city that the post be transplanted to the median on Sixth East "in order that the relic may be preserved for sightseers." That is apparently what happened. Somewhere along the way, though—by at least 1933, when the DUP erected the marker—it became the Lone Cedar Tree, so christened by the unreliable but profound forces of community memory.

A juniper clings to a cliff on Cedar Mountain, Utah.

3

Survival

Late on a July evening, Eddie and I drove up through the mountains west of Utah Lake. Yellow light spread over innumerable cedars. Sagebrush glowed. The rough road wound through steep and folded hills, the same route that U.S. soldiers of the 1850s used as they traveled to and from their base at Camp Floyd in Utah's West Desert. The soldiers had marched to Utah Territory under orders from the president to replace Brigham Young as territorial governor and keep an eye on the unpredictable Mormons. The Mormons prepared for war, but in the end the new governor took over peaceably.

Several years earlier, Eddie's friend and colleague, flying an Apache helicopter, had crashed into a juniper woodland up in these hills. The tragedy had haunted Eddie since then, and he felt he had to find that crash site. We had come into this fading light with the coordinates and a borrowed GPS unit to look for it.

Juniper or pinyon-juniper woodlands occupy millions of acres in Arizona, Nevada, New Mexico, and Utah—growing on as much as 21 percent of the total land area in Utah and 17 percent in New Mexico. Utah juniper is the most common tree in the Great Basin. It also takes root in Colorado, southern Idaho and Montana, southeastern California, and western Wyoming. It makes a life in many different types of ecosystems: ponderosa pine, western hardwood, sagebrush, desert shrub, chaparral–mountain shrub, mountain grassland, desert grassland, and of course the pinyon-juniper ecosystem.[1]

In the Lake Mountains, junipers grow interspersed with a rocky, dry sagebrush community. On that July night, we parked and then walked through a twisting of trees and earth, veering and correcting our course down a slope, across a gully, searching for the signs. A nighthawk flew overhead. We reached a spot where broken juniper branches were scattered along a gentle slope. Woodlands stood as witnesses around a scar in the earth, a bare spot. Old, old trees stood around this place, on soils and rocks where not much else grew. The quiet of crickets sang the twilight's fading. I imagined a helicopter coming in fast and low, blades beating back trees, lives shattering.

Eddie had to ponder a while. He adjusted a dead cedar branch and readjusted it, perhaps looking for pieces of memory. One big tree had a broken top. Maybe the blades had clipped it, and the 'copter spun into the hill. If the pilots had been only fifteen feet higher, or if there had been no trees.... But there were, there are. These slopes in the Great Basin harbor junipers. They are just right for junipers. As we silently picked our way back through the trees, dusk became night.

*

The same thing that discourages many species helps make junipers common in the West: aridity. The junipers of the western United States not only survive but thrive where little moisture falls. Utah juniper flourishes in areas where twelve to eighteen inches of precipitation fall annually, but it can also live quite well on less than ten. Not every plant can prosper where summers are hot and dry and winters are cold and swept with wind and snow. Not many plants can spring up and endure on alluvial fans, on dry, rocky hillsides, and in shallow, alkaline soils. But in the West the juniper takes on these challenges and can elbow out other plants that might be willing and able to grow in these conditions. Really dry conditions may slow or stunt juniper growth, though; one writer even claims, incredibly, that in severe climates a Utah juniper might grow for fifty years and get to be only six inches tall.[2]

There's one species of juniper in the West that wouldn't do so well in a climate that dry. Rocky Mountain juniper needs higher elevations and latitudes than its more drought-tolerant cousins do. This tree grows all the way into Canada and also into West Texas and as far east as western Nebraska, often hanging out with trees that like cooler, moister places—trees such as Douglas-fir, larch, lodgepole, and cottonwood. It can also live comfortably with sagebrush, prairie grasses, and shrubs. And in some places it grows alongside various other juniper and pinyon species in pinyon-juniper woodlands—but only when those places are not severely dry.[3]

Because junipers of different species do thrive across the North American West and cover so much of an otherwise austere landscape, they have intertwined with the doings of humans since humans first came here. Explorers and emigrants often mentioned cedar trees, and often these trees became part of the unfolding stories of their travels. In 1825, William Ashley, the man who started the far-reaching Rocky Mountain Fur Company, journeyed down the Green River in a bull boat. This he and some of his trappers built by stretching buffalo hides over a willow frame and then waterproofing the whole thing with pitch and tallow. Though not exactly a craft for big rapids, it was a resourceful one. With it, Ashley and his men floated and portaged from southeastern Wyoming through Flaming Gorge, Browns Park, and Lodore Canyon, and on to the start of Desolation Canyon in Utah. He was looking for an easy way to transport furs. The Green, with its rapids and, not incidentally, the wrong destination, proved not to be that easy way. But the trip increased Anglo knowledge of the West. Within the deep walls of Lodore Canyon, Ashley and his men encountered a winter encampment of "some thousands of Indians" in the willows along the river. Ashley noted that the people covered their lodges "with the bark of cedar." In this location in mountain canyons, the cedar bark could have come from Utah junipers or Rocky Mountain junipers, those trees that need a bit more moisture than some other junipers do.[4]

A little shift of some kind, and organisms diverge in subtle ways, or in striking ways. One goes in one direction; another goes in a different direction. Trees—individual trees as well as tribes of trees—adapt to their own niches, as do persons, families, and cultures. Circumstances, environment, and other beings influence us all. Rocky Mountain juniper (*J. scopulorum*) differs from other western junipers because it followed a different evolutionary path. It lives in the West but actually evolved as part of what is now an eastern U.S. group (or "clade") of junipers. This group includes eastern red cedar (*J. virginiana*) and junipers of the Caribbean, and the members all require more moisture than those of the western clade. So, although they can live near each other sometimes, Rocky Mountain juniper and Utah juniper aren't the closest of cousins.[5]

Members of the western clade of junipers range from northern Guatemala to Canada, and from the far western states to as far east as Arkansas. This wide-ranging clade may have originated in north-central Mexico, because no other region in the Western Hemisphere has a larger number of *Juniperus* species. Over millennia the juniper ancestors acquired tricks for living sparsely and for thriving in parched circumstances, and they passed on that quality to grandchildren. So now their descendants are "xerophytic," able to live in deserts and arid high country.[6]

*

This very trait enabled a couple of junipers to save the lives of men crossing one of these deserts—the Great Basin. The very intrepid—indeed, seemingly fearless—Jedediah Smith made the first attempt that we know about, crossing the dry Basin and Range landscape of Nevada and Utah in 1827. Smith had come West in answer to William Ashley's ad for "one hundred enterprising young men" to join his fur company. He aimed to earn money for his family and, perhaps most compellingly, to satisfy his large hunger for adventure. Among the mountain men, Smith stood out. He seems to have been quiet, modest, and intelligent. He carried a Bible and read it. He was cool headed: when a grizzly bear pretty much ripped off his scalp and ear, he asked James Clyman to sew it back on. He was selfless: he sent his brother $2,200 (perhaps $40,000 to $50,000 in today's dollars) to take care of their aging parents, writing, "It is that I may be able to help those who stand in need that I face every danger.... Let it be the greatest pleasure that we can enjoy to smooth the pillow of [our parents'] age and as much as in us lies, take from them all cause of trouble."[7]

And of course, he was audacious: in 1826 he set off to explore from Utah to California and back, because "I wanted to be the first to view a country on which the eyes of a white man had never gazed." He led his group southward from Bear Lake, on the border of Utah and Idaho, through high and low deserts and over mountains into California and on to San Diego, then up to the American River. This was no Sunday stroll, but the return trip in 1827 would tax him to his limits. He took only two men and began by crossing the Sierra Nevada. Five hundred miles of the Great Basin lay ahead, and the month was

June. The three pushed through as quickly as possible, traveling as much as forty miles a day on little water and not much meat—they ate horse flesh some of the time.

The group missed some of the few springs along the route and began to suffer sorely from lack of water. On June 25, Smith climbed a hill to see whether he could spot vegetation that might indicate moisture. He could see nothing but austere land and a snowy mountain perhaps sixty miles distant. "When I came down I durst not tell my men of the desolate prospect ahead," he wrote in his journal. Instead, he tried to instill hope. Urged on by Smith's optimism, they "pushed forward . . . over the soft sand. That kind of traveling is verry [*sic*] tiresome to men in good health who can eat when and what they choose, and drink as often as they desire, and to us, worn down with hunger and fatigue and burning with thirst increased by the blazing sands, it was almost insurportable [*sic*]."

And here, a "small cedar" comes into the story. At four in the afternoon the three stopped to find relief in its shade. "We dug holes in the sand and laid down in them for the purpose of cooling our heated bodies." After resting, they got up and pushed on. Smith thought, "How trifling were all those things that hold such an absolute sway over the busy and prosperous world. My dreams were not of Gold or ambitious honors, but of my distant, quiet home, of murmuring brooks, of Cooling Cascades."

They staggered through the night. In the morning, the hot landscape again tormented them. "At 10 O Clock Robert Evans laid down in the plain under the shade of [another] small cedar, being able to proceed no further." If he was to live, it was up to Smith and Silas Goble to save him.

Three miles later, "to our inexpressible joy," Smith and Goble found water at the base of the mountain. Smith took some back to Evans. "He was indeed far gone. . . . Putting the kettle to his mouth he did not take it away until he had drank [*sic*] all the water, of which there was at least 4 or 5 quarts, and then asked me why I had not brought more."

Two days later, the three joyfully sighted the Great Salt Lake—"my home of the wilderness," Smith called it.[8]

*

Junipers have a lot of strategies for surviving in such parched landscapes: the robust taproot that dives down to deep water sources and the long lateral roots that suck up any moisture that seeps down from the surface; resinous, thick, overlapping leaves that hold on to water; a thick litter of leaves and seeds on the ground that keep out competition and help slow evaporation; seeds that lie dormant in times of drought; yearly changes in the ratio of female to male cones; light-colored foliage that reflects sunlight; the tendency to stop growing during dry times. All of these add up to a plant that can tough out some pretty serious droughts.

Lots of plants do similar things. But junipers have another, less visible, strategy—one that gives them an edge over other trees, like pinyons, for instance. Pinyon-juniper woodlands are iconic landscapes in the West because these two trees flourish in similar conditions. But not always. When things get hotter and drier, pinyons drop out. Pinyons are not exactly pampered hothouse plants, but severe droughts can wreak havoc on pinyon populations while junipers may survive and spread. Why? Junipers can resist "cavitation" better than pinyons can.[9] Cavitation—a disruption in the flow of water—kills trees. It happens like this: as water evaporates from a tree's leaves, a vacuum-type effect sucks water through the xylem, a thick layer of wood between the heartwood and cambium. Normally, a constant tension keeps the column of water moving up the xylem, but in dry conditions, when water is harder to get, that tension grows. Like a rubber band being stretched and stretched, the column can snap. Air bubbles can then move into the gap, especially when the xylem itself has ruptured.[10] That's cavitation. Suddenly, part of the tree or the whole tree has lost its water source and is at risk of dying.

A study of fourteen juniper species found that those that evolved in dry conditions—the western clade—can resist cavitation better than their eastern clade cousins. As they evolved, the western junipers added extra woody tissue to reinforce and strengthen the xylem, helping prevent ruptures of this layer. Not incidentally, the champion cavitation resister—*Juniperus californica*—grows in the Mojave Desert.[11] "The take-home message is that junipers are the most drought-resistant group that has ever been studied," says Robert Jackson, an author of the study.[12]

All of this means that when people are in unsentimental, harsh landscapes, junipers are often there, too. It was beneath a juniper on Island in the Sky mesa, high above the confluence of the Colorado and Green Rivers, that a man sat down and died. Edward Abbey joined the party searching for the man in the furnace-like heat of summer. They found him bloated and dark there in the hot shade.[13]

*

Junipers can speak to the extremities of a moment. They framed an event in 1846 when the Bryant-Russell Party was heading to California. Like Jedediah Smith, this group was about to cross the Great Basin, only in the other direction. Edwin Bryant had left his life as a newspaper editor in Kentucky to journey across the continent and, like a true journalist, write a book about it. His book about the trail and the golden place at trail's end would be lively, full of stories and descriptions, and eagerly read by many.

But first he had to get to California. He and eight other men traveling on mules had journeyed with the Donner Party for much of the way from Missouri. Finally, nervous about the slow pace of the Donner group's heavy wagons, the Bryant-Russell group split off on their own. They were following a supposedly "faster" and "better" route that Lansford Hastings and James Hudspeth had been promoting as an alternative to the Oregon Trail. While the Oregon Trail crossed southern Idaho, the new route crossed the Wasatch Range and then skirted the southern end of the Great Salt Lake and pushed on across the desert. The men on mules quickly realized that this would be no route for wagons. Bryant claims that they sent back word to Fort Bridger that the Donner Party must *not* come this way.[14] Very unfortunately, the message did not reach the Donner group. Their difficulties in cutting a road over the mountains and dragging their wagons through the salt-crusted muck of the desert delayed them, and thus they got caught in the snows of the Sierra Nevada.

Almost four weeks ahead of the Donner Party, Bryant and his fellow travelers camped on August 2 by a ravine where the sand was damp, on the east slope of the Cedar Mountains.[15] On the other side of the mountains lay the Great Salt Lake Desert. The next day the group would try to cross this desert. At camp, they dug into the sand to get a little salty, sulfurous water—the last they would find until they reached the base of Pilot Peak at the western edge of the desert. As Bryant later wrote, he woke at 1:30 a.m. with feelings of

A juniper on Cedar Mountain overlooks the Great Salt Lake Desert.

apprehension. The surroundings, including the junipers, mirrored his feelings: in the moonlight "all was silence and death. . . . No rustling zephyr swept over the scant dead grass, nor disturbed the crumbling leaves of the gnarled and stunted cedars, which seemed to draw a precarious existence from the small patch of damp earth surrounding us." Later that morning, as the men picked their way down the western slope of the mountains, a few "straggling, stunted, and tempest-bowed cedars" would seem like fellow strugglers. But now, in the predawn hours, the men gathered dead limbs of juniper that, they believed,

John C. Frémont's party had cut when they had camped here during their 1845 expedition. In the light of cedar fire they ate a frugal breakfast—forgoing bacon, to avoid aggravating their thirst—packed the mules, and made their plans.

Mountain man James Hudspeth, who had guided them to this point, led them for a way and then urged them, "Now, boys, put spurs to your mules and ride like hell!" And they did. At least, they rode off at a brisk trot. They pushed through salt, sand, and mud, a fierce windy salt storm, mirages, and burning thirst. They dismounted and walked or staggered when the mules got too exhausted. Late at night, the nine men reached the life-giving spring at Pilot Peak, having crossed seventy-five miles of salty muck in one day by traveling light and fast with their mules.[16]

Bryant went on to become the alcalde (or mayor) of San Francisco for about five months, sell thousands of copies of his book, make a bundle in San Francisco real estate, build a mansion in Kentucky, struggle with poor health, and, in the end, jump from a window to end his life.

*

A couple of years later, a far different party was making a different journey through the West, not through summer's heat but winter's cold. In December 1849, this exploring expedition set off from Great Salt Lake City southward to take a broader look at the Mormons' new homeland and find good sites for settlements. Parley Pratt, Mormon theologian, missionary, apostle, road builder, and more, had charge of the fifty-man "Southern Exploring Company." The official clerk of the company, twenty-four-year-old Robert Campbell, kept a detailed journal. During the dark winter months, the men explored down the route of Highway 89 until they turned west to cross the Tushar Mountains. They then explored as far south as present Saint George and back up to Salt Lake City along the future route of Interstate 15, returning home in March.

Like diarists before him, Robert Campbell noted the landscape for those to come, and he wrote of junipers growing across the frozen terrain:

> December 12, camped on the Sevier River in Marysvale Canyon: "cedars plenty in creek & on the sides of the Mts."

> December 17, crossing the Tushar Mountains south of Circleville: to get up and down the steep Tushars, men moved rocks, leveled steep places, and locked the wheels to go downhill, pulling backward on ropes to slow the wagons. "Boys cold & tired, snow so deep to wade thro holding wagons back, pulling oxen up steep pitches then oxen pulling the wagons, tedious work. Wind, & small frozen flakes of Snow on the tops of the ridges like to blow thro a fellow." The labor was made harder because, as Campbell wrote, "Plenty cedars" covered the slopes.
>
> December 20, at 7,500 feet elevation: "The land now hilly & knolly a head S & W many miles not even cedar timber. . . . 8 or 10 miles a head the bench land . . . covered with tender cedar."
>
> December 25, letter from Campbell to LDS First Presidency Brigham Young, Heber Kimball, and Willard Richards: "The mountains and some bench lands where we have passed thro is thickly studded with cedars."

The men had reasons for all this attention to junipers. As Pratt explained in his official report, these trees could contribute to the settlement effort. "But the best of all remains to be told," Pratt wrote. "Near the large body of good land [Cedar Valley] on the Southwestern borders are thousands of acres of cedar contributing an almost inexhaustible supply of fuel which makes excellent coal. In the centre of these forests rises a hill of the richest Iron ore."[17]

He meant that the junipers would supply charcoal, not coal, of course. And he meant that Mormons should mine and, with that charcoal, smelt the ore they had found—rich enough that the county would be named after the iron.

*

The juniper woodlands that crowd the arid American West have thus played many roles in human struggles to live and prosper. In 1865 they played a much different part. Around this time a Ute whom the Mormons called Black Hawk decided he must do something about the Euro-American invasion of Ute lands. He began to attack and raid settlements. A loose coalition of Utes, Paiutes, and Navajos joined him in this last-ditch effort to regain what they had lost. The conflicts ranged from cattle-stealing raids, to brief skirmishes, to battles, to massacres by both sides.

In early summer 1865, Captain Jesse West marched a group of militiamen to Thistle Valley in Sanpete County to protect the settlers there. The men camped at the edge of a juniper woodland. On the morning of June 24, West later said, he heard a rifle shot and saw two of his men "just emerging from the cedars." Five Indians were galloping toward them, shooting. Charles Brown fell, mortally wounded. "As he lay dying, one of the guards, a young private named Garr, knelt down, took careful aim, and fired. His shot scattered the horsemen, who retreated to the cedars in haste, and the body of Brown was brought into camp." Brown and another man had disregarded orders to stay in camp and had gone out into the trees. "Suspecting no danger, they were busy picking gum when the shot was fired which cost poor Brown his life."

After that, the battle heated up with intense fire; West felt that the Utes had the advantage, since the thick cedars hid and protected them. Toward evening, the Indians advanced closer to the camp, crouching in some brush near the creek and firing at the exposed camp of the militia. Bullets whizzed through the air. The situation had become desperate, but to the militia's relief, reinforcements suddenly appeared. The Indians retreated to the cover of the cedars, and the battle ended.[18]

*

Cows found protection in juniper woodlands too. In southeastern Utah, cattle left on the open range tended to become wild and skittish. When the cowboys came to round them up, they ran, dodging at full speed through a wild terrain: a jumble of gullies, canyons, and cliffs, and everywhere thick stands of pinyon and juniper. A cowboy didn't have the leisure to wind his way around the trees and across the washes and gullies. To catch the cows, he'd have to run his pony full speed, dodging the snaggled limbs of juniper and pinyon by jerking his body from side to side or ducking. If he didn't have quick enough reflexes, he could get knocked off his horse or skewered. These were juniper encounters of the most literal kind. Charles Redd, who later became a respected rancher and a leader in civic affairs, remembered his days chasing those cows: "Just catching up with the wild cattle was a sort of victory all in itself. For a cowboy couldn't begin to dodge all the trees in his way. He had to hit many of them and hit them so hard that he'd break the limbs. And both horse and rider had to have pretty fair judgment about how big a tree and how big a limb would break. But they could never hesitate. They had to hit it so hard that something would give, and they were always in hopes it would be the tree or limb, not them."[19]

*

In all these juniper encounters in desert country, a person's experience of the tree depended on what he or she needed or was trying to do. The very young artist-wanderer-dreamer Everett Ruess saw them with an artist's eye. As he wrote to his parents: "I enjoyed riding down from Bryce canyon through the grotesque and colorful formation. Mother would surely enjoy the trees; they are fascinating, especially the twisted little pines and junipers." Ruess's eye for the shape and soul of juniper and all wildness lives on in the block prints he created. He was only twenty when, in 1934, he and his burros walked out of the tiny town of Escalante, Utah, and vanished into legend.[20]

In 1907 certain land promoters saw junipers with eyes fixed less on beauty than on the bottom line. The real estate they were marketing in the Uinta Basin was a tough sell, but a true promoter can find a way to use even junipers to advantage. The company took photographs of the land showing abundant trees, and in the black-and-white photos the trees look appealing and lush. The promoters then wrote copy for their brochure implying that the area was perfect for growing fruit. Lured by images of junipers that they took to be orchards, some people hurried to get in on the deal. But of course, their orchard-growing dreams withered as soon as they realized that at nearly seven thousand feet, with poor soil and little rain, this land was anything but a fruit paradise.

Nevertheless, that's how the place—originally known as Rabbit Gulch—took on the name Fruitland. And in arid Fruitland, magnificent junipers still grow in abundance.[21]

4

Spirals

By trail, it's five miles uphill to reach the biggest Rocky Mountain juniper in the world, up Logan Canyon in northern Utah. It's a fine trip, kind of tiring, but if you haul yourself up there on a bike or skis, the downhill is bliss.

Pedaling up that trail years ago in June, we passed aspen, silvery sagebrush, blue flax, geranium, mules ear, and waving grasses. High up, a twisted old juniper along the trail stopped me short. This wasn't the world champion juniper, just an obscure old tree, but I couldn't stop looking. I lay down beneath it and just gazed at its ridged skin and the ragged bits of bark, the twists and intricate curled branches. These coiled, spiraling old branches fascinated me, as if they held some mystical secret. When did that curve appear? How long had that branch been splaying its leaves into the sun?

I felt the kind of reverence I have felt before a van Gogh. The tree was swirling yet still motion. Its massive roots curled around rocks. I could almost feel the slow, sweet movement of sap, the life force flowing up and down. The living wood beneath the ancient skin nourished twigs holding up clusters of scaly leaves that swayed in the wind.

Even though wildly spiraling trunks and branches look impressive, spiraling trees are not really unusual. They're just not usually so extravagant. You can see spiral grain fairly often in debarked trees, old wooden telephone and power line poles, and pilings. You might detect it in spiraling bark cracks.[1] According to a log home builder, the logs he works with almost always have some spiraling.[2] If you could look underneath the bark of a young tree you might see the grain curving gently upward and leftward. (Leftward means that if you put your

The twists and curves of one juniper trunk.

right hand on the tree trunk, fingers up, your thumb would be pointing in the direction of the spiral.) Trees tend to begin spiraling in this leftward way, and some trees may continue in this direction. If they do, the wood will be weaker than it could be. But in other trees, as they mature, the leftward spiral stops. The tree might grow straight for a few years and then begin to spiral to the right. This change of direction strengthens the wood.[3]

When you see an old dead juniper snag and the curve of its gray-weathered grain, you are seeing the basic structure of the tree. You can see how the tree negotiated time and space, the directions the cells took as they divided, again and again. Sometimes you see the wood spiraling right, sometimes left, and sometimes not at all.[4] In the early 1900s, a man named Theodore Cook became obsessed with spirals he noticed in nature and studied them extensively. After all his reflection, he came to believe that "with very few exceptions the spiral formation is intimately connected with the phenomena of life and growth."[5]

*

When we finally reached the grand old lady, the Jardine Juniper, we stood in amazement. No sculpture by anyone anywhere can match this balance of light and dark, smooth and furrowed, up and around. The tree, around forty feet tall, leans northward. South-leaning branches give counterweight. One dead branch twists around and around. The dead trunk spirals upward also. It holds dead and barren branches, limbs, fingers, and a small crown of green. Just that. A strip of life within this shell of mostly lifeless wood nourishes the green.

A nearly illegible old wooden sign said, in all caps: *JARDINE JUNIPER. THIS ROCKY MOUNTAIN JUNIPER HAS BEEN ALIVE FOR 3,200 YEARS. WITH RESPECT IT WILL SURVIVE MANY MORE. PLEASE ENJOY THE TREE FROM A DISTANCE.* The experts say that the tree is probably half that old, and in fact when I returned a few years later, the sign had been changed to read "1,500 years." But even so, it would have been a baby when the legendary King Arthur was fighting the Saxons. The tree grows on a limestone outcropping splashed with orange lichen. The Forest Service had put a couple of notebooks in a box for people to sign. A few of the remarks:

The Jardine Juniper, clinging to life on a mountainside in Logan Canyon.

"So peaceful. It feels like there is no evil in the world!!!"

"Well, tree, this time it's just me. I gave you some of Zeke's ashes to help you grow for another thousand years or so. Zach is old & slow & the trail is long, so I didn't drag her up here. . . . Live long & prosper, great tree!"

"Freakin Sweet!"

"This tree has 150 lifetimes to my one. I suspect it has collected rather more wisdom than I have. I'll take away what I can. Also I notice a chipmunk has its home amongst the roots. Remember, the sacred is part of you day to day."

"These are the times I live for."

"Need to make firewood."

"She is a grand, tough old broad. I hope I look like she does when I'm hundreds of years old, because seriously, women in my family DO NOT age well."

"Survived 21 hours of air travel with 3 connections to get to SLC so I could drive here and hike. It was completely worth it!"

"The Jardine Juniper is Logan's 'Tree of Life.' Reminds me why life is so precious, hikes like this you never want to end."

"Well it's been a hell of a ride."

"Very gnarly tree yo!"

While I sat there and Eddie wandered around taking pictures and startling a huge rattlesnake, some people hiked in, a man, woman, two kids, and two dogs. "Geez, that's crazy!" "Look at the branches!" "Wow." The man read the sign: "Three thousand, two hundred years, James." He told his son not to climb the tree.

*

What makes a particular tree spiral at all? Why not just grow straight?

The spiraling, of course, begins at the cellular level, resulting from the division of cambium cells. According to joint research by a botanist and a mechanical engineer, as cells divide "there is a constant, incessant tendency of all maturing cells to change their orientation in a given direction."[6] A cell tends to orient in the same direction as the cell that preceded it, but what makes the cells start growing spirally in the first place?

Twisting tree in Utah's West Desert.

Those who have thought about this have suggested various possible causes:

Physical stress on the cells, such as prevailing winds nudging the tree year after year. This theory has been around for a while: Theophrastus—student of Aristotle and "father of botany"—wrote two thousand years ago, "But if a tree stands sideways to the north with a draught round it, the north wind by degrees twists and contorts it, so that its core becomes twisted instead of running straight."[7]

Snow loading, or harsh environments in general.

A particularly fierce storm when a tree is young.

The angle of the slope a tree is growing on.

The Coriolis effect of the earth's rotation (the same force that causes things like hurricane spirals).

An adaptation arising from the need to deliver water and nutrients more evenly and effectively—especially when water is not available to all the roots.

An interruption in the conduit for water (the xylem), causing new cells to bypass the break.[8]

Genetic propensity, possibly triggered by stress hormones.

The log home builder gives his take: "All the science I have read agrees: spiral grain is overwhelmingly genetic. The seeds and cones from righthand trees tend to produce righthand young'ns. It is not surprising that most of the trees on a north slope somewhere have about the same spiral grain—they're closely related.... I guess cones don't fall far from the tree."[9]

If you contemplate it with a nonliteral mind, twisting wood may evoke larger phenomena, like the spiraling nature of life itself. John Burroughs wrote: "Nature does not balance her books in a day or in ten thousand days, but some sort of balance is kept in the course of ages, else life would not be here. Disruption and decay bring about their opposites. Conflicting forces get adjusted and peace reigns.... There is a perpetual see-saw everywhere, and this means life and motion."[10]

Look, and see the spiral all around: whirlpools and hurricanes, tendrils of peas, seeds in a sunflower head, snails, rose hearts, fiddleheads on ferns, ram horns, embryos. Ancients all across the planet painted spirals and pecked them into rock, sometimes in connection with sites aligned with the solstice. Today, we speculate that these prehistoric spirals point to the circular nature of time, or of the migrations of people. And so it is: time cycles in rhythms small and large, from the rising of each day's sun to the slow spinning of galaxies, stars being born and stars dying. On earth, the geologic cycles repeat over and over: sediment beneath an ocean that becomes rock that becomes mountain that becomes sediment running down to an ocean, over and over. A spiral always grows but never covers the same ground, illuminating the past and pointing to the future.[11]

As the stars, the earth, and all creatures do, the juniper has its own deep, cyclical history.

Imagine that for the last forty thousand years you have hovered high above the American Southwest—like an immortal hawk coasting on stratospheric thermals—watching. Over the centuries you would have seen shifting mosaics of woodlands, grasslands, forests, deserts. As the earth cooled and warmed, as it grew wetter, drier, and then wetter again, vegetation would adapt and migrate, shifting to new places and then back again. Fires would burn large swaths into the forests. Afterward, plant life would return and over time reestablish the "status quo"—until the next fire. Over wide cycles of time and climate you would see junipers move to different elevations and latitudes, disappear from these same places, and then return again.

After humans moved into this landscape some twelve thousand years ago, you would see their impacts as they set fires or cut trees for firewood and structures. Over time, as human populations and their technology grew, you would see larger impacts: more trees cut for fences, mining, and firewood. Humans would also bring in cattle and sheep that decimated grasses. They would increase the carbon dioxide in the air. Under these conditions, coupled with a warmer and drier climate, new trees would sprout in different places and grow into dense woodlands. Humans would sometimes respond by pushing the new woodlands back. All these "little" acts of humans would be playing out within the large, repetitive patterns of shifting climate.

So, as you hovered above the earth, you would see the deep spiraling story of junipers played out in motion across the landscape. You would see births and deaths, generations rising and falling. If we ever had the illusion that anything could be permanent, watching through the millennia would certainly set our view straight. If we ever had the illusion that humans are lords of the landscape, watching would cure us.

Actually, paleoecologists *do* see these changing cycles of landscape, not from the air, but in the field and in their labs. As they research ancient ecological systems, paleoecologists can see the long view in records laid down decade by decade in lakes, wet meadows, and bogs: charcoal, pollen, insect remains, and vegetation stratigraphies.[12] Tree rings from both living and long-dead trees also help weave together a view of the climate cycles.

And then there are woodrat middens—heaps of hoardings that can be astonishingly old. Woodrats—also known as packrats, or in scientific classification *Neotoma* spp.—live in caves and alcoves or around old stumps or trees, and they collect stuff. They collect and pile up leaves, grasses, seeds, twigs, and small bones, and then they pass the job to their descendants, for centuries. For millennia, sometimes. One desert woodrat nest in the La Sal Mountains produced ten bushels of materials: 80 percent sticks and twigs, 5 percent empty pinyon cones, 4 percent bones, 2 percent rocks, 2 percent deer hide and hair, 1 percent cactus, 1 percent mushroom bits, plus a quarter pound of pinyon nuts and a small amount of juniper berries.[13]

The woodrats do us a favor not only by collecting a remarkable record of the past but also by urinating on that record through the years. The urine crystallizes and cements the pile into a hard mass. A midden in an alcove or cave is thus preserved for the ages, as long as a predator looking for a woodrat lunch doesn't break in. One large midden collected by generations of enterprising woodrats—but perhaps from a radius of only three hundred feet—can carry tens of thousands of years of ecosystem information. Fortunately, rock shelters and the West's arid climate have helped some of these valuable "libraries" survive that long.

An investigation of twenty-two packrat middens in the Little Nankoweap drainage of the Grand Canyon turned up plant remnants from almost 46,000 years before the present (BP). Most of these middens were around 3,800 feet in elevation, on ledges and in caves in steep cliffs. Today, only plants like yucca, cactus, globemallow, and sparse Indian ricegrass grow there. But between 10,000 and 50,000 years BP, juniper was a constant presence even as climate and other vegetation shifted. Around 17,000 years BP, Utah juniper grew on the ledges of Little Nankoweap alongside Rocky Mountain maple, Douglas-fir, limber pine, orange gooseberry, and common snowberry. These past companions now grow at much higher elevations, but for 40,000 years they migrated downslope during cool, wet periods and upslope when the climate warmed up. Utah juniper persisted there when other trees could not, however, becoming most abundant between 20,000 and 10,000 years BP.[14]

Likewise, Utah juniper has flourished in the Great Basin for at least 35,000 years, growing widely over space and time, through dramatic shifts in climate. During that time other plants came and went, including one-seed juniper, but Utah juniper and a handful of other plants, such as sagebrush, ephedra, and brickellbush, persisted. Still, as the climate warmed up after the last Ice Age, even Utah juniper lost its ability to survive at elevations below four thousand feet.[15]

*

House Rock Valley straddles the Utah-Arizona border, just north of the Kaibab Plateau. The Vermilion Cliffs tower up along the valley's edge, and here is where, starting in 1996, California condors have been regularly released into the wild. Now, some seventy-five condors soar with nine-foot-long wingspans above House Rock Valley and Vermilion Cliffs National Monument—a pretty good group, since only twenty-two living condors were known in 1982, and today fewer than 250 total condors are flying in the wild.[16] Below their wings, much of House Rock Valley now harbors only sparse vegetation. However, if you extrapolate what the woodrat midden data from Little Nankoweap drainage are suggesting, you could surmise that at one time a Douglas-fir–Utah juniper woodland flourished in this now-desert valley.[17] Now, in the twenty-first century, there is not a Douglas-fir in sight, and even the Utah junipers in House Rock Valley are struggling. Many of these old-growth trees harbor masses of tangled yellow-green "twigs"—mistletoe. These parasites send roots into the phloem and xylem of trees and suck out water and nutrients.[18]

On a day in May, rancher Steve Rich gunned his old Tahoe across sandy washes, bounced over rocky "stairs," and swept along the twisting turns of a House Rock Valley side road. These roads severely challenge most drivers, and Rich makes it very clear that no one should take a lightweight four-wheel drive back in here. The junipers grow so thickly along the roads that you can't avoid getting your vehicle scratched up. Mistletoe has affected the woodlands, and many trees are sickly or dying. Rich says that the trees wouldn't be so heavily infected if they were growing in their ideal habitat. These trees in House Rock Valley are now on the very edge of the habitat that suits them best. The climate stresses them—and as many of us know firsthand, stress can certainly open up an organism to parasites and disease. For instance, in one study, junipers growing in poor soil had more mistletoe freeloaders than junipers growing nearby in better soil.[19] When the centuries-old trees in House Rock Valley first sprouted, the climate may have suited them fine, but now, along with all species, the junipers are having to deal with a warming climate. Here they seem to be in the slow process of dying out and "migrating" upslope to conditions that suit them better.

During the last major ice age, chaparral and juniper-dominated woodlands actually prospered at ridiculously low latitudes and elevations, such as the Mojave, Sonoran, and Chihuahuan Deserts—places where they could never grow today. At the same time, alpine trees grew on the now-dry and hot foothills of the Great Basin, while snow smothered the mountains, and lakes and wetlands filled the valleys. Around 11,500 years ago, the snows began to recede, the climate began to warm, and the conifers moved upslope. Pinyons disappeared from the low desert elevations, while junipers and scrub oak kept hanging on there for another two thousand to three thousand years. But gradually the deserts grew too hot and dry for even these tough trees, and juniper generations migrated northward and upward. The climate has cycled a few times since then. Woodrat middens tell the story.[20]

At this moment in the earth's deep history, Utah juniper grows as far north as southern Montana, at East Pryor Mountain and Bighorn Canyon. According to midden evidence, the tree arrived at Bighorn Canyon more than 4,600 years ago, perhaps just as the Rocky Mountain juniper and creeping juniper that had earlier dominated the canyon were on their way out. Not until 3,000 years later did Utah juniper make it to a spot only thirty kilometers away—the western slope of the Bighorn Mountains and the Wind River Canyon in Wyoming.[21] So, Utah junipers migrated into central and northern Wyoming and southern Montana from the south slowly and seemingly randomly in a "series of long-distance dispersal events"—sometimes jumping as far as 135 kilometers to the next suitable site.[22]

Think about this for a minute. How and why did juniper seeds cross those distances? There are many interconnections at play here: the lay of the land, a particular year's weather, the climate in general, the presence and doings of animals and humans, the flow of water, and probably more. Juniper migration is a small topic in the large scheme of things. But in knowing the outlines of the juniper's deep history, I can glimpse the spiraling of events that have led to the current vegetation structure in the Wasatch Range near my home, or anywhere. Plants have been evolving, migrating, expanding, and contracting for as long as plants have existed.

*

An extravagantly spiraling branch of the Jardine Juniper.

Smithson's *Spiral Jetty* in the "immense roundness" and "gyrating space" of the Great Salt Lake.

The spirals of twisted trees like the Jardine Juniper—and spirals throughout nature—endlessly fascinate. From Stone Age times to now, we seem drawn to spirals as a species, looking for them, creating them, finding meanings in this ever-repeating, ever-changing pattern. Robert Smithson felt compelled to make a gigantic earth spiral, inspired by the land itself. When he first came to the site on the Great Salt Lake where he would create the *Spiral Jetty*, he felt himself in the center of a circling landscape. He saw irregular beds of limestone, massive deposits of broken black basalt "giving the region a shattered appearance," mud cracks under the shallow pinkish water. "As I looked at the site, it reverberated out to the horizons only to suggest an immobile cyclone while flickering light made the entire landscape appear to quake. A dormant earthquake spread into the fluttering stillness, into a spinning sensation without movement. This site was a rotary that enclosed itself in an immense roundness. From that gyrating space emerged the possibility of the Spiral Jetty."

Smithson saw in his jetty a continuation of the earth's spirals. "Each cubic salt crystal echoes the Spiral Jetty in terms of the crystal's molecular lattice," he wrote. "Growth in a crystal advances around a dislocation point, in the manner of a screw. The Spiral Jetty could be considered one layer within the spiraling crystal lattice, magnified trillions of times."[23]

My father used to quote Oliver Wendell Holmes's "The Chambered Nautilus," which he had to memorize in school. In the poem, spiraling is a metaphor of ever-onward growth: the nautilus *becomes*, chamber by chamber.

> Year after year [he] beheld the silent toil
> That spread his lustrous coil;
> Still, as the spiral grew,
> He left the past year's dwelling for the new,
> Stole with soft step its shining archway through,
> Built up its idle door,
> Stretched in his last-found home, and knew the old no more.

*

As with the chambered nautilus, the present state of an ecosystem arose from the past and contains the past. The long cycles of geologic processes, the smaller cycles of climate and vegetation changes, and the relatively tiny cycles of human influence are all imprinted onto the land today. In particular, major disturbances leave their mark. When we walk through a forest today, we walk through vegetation heavily influenced by changes in the past.[24]

The vegetation responds to an upset to its "equilibrium"—which is always tenuous anyway—and may recover, reestablishing some sort of balance, but it is changed. History makes us familiar with the process on a human scale, if we haven't experienced it ourselves. Something drastic happens. War, disease, earthquake, famine, environmental destruction, changing climate—whatever it is. People die, are injured or sickened. Over time, groups regroup. People find new ways to live, the way wood may spiral after a storm or damage. The disaster becomes ingrained in the culture and psyche, through traditions, stories, ways of doing things. For the survivors, life goes on, changed. We are who we are because of past events that we may have forgotten now. So it is with vegetation.

In many cases, humans *are* the disturbance. For instance, over time, the people living at a major prehistoric population center in Chaco Canyon, New Mexico, cut down most of the surrounding trees for fuel and buildings. The Ancient Puebloans here and elsewhere liked to use juniper for roof beams, probably because it is less susceptible to boring insects than is pinyon wood.[25] In fact, the people valued and reused old juniper beams for new construction (which particularly makes sense if your cutting tool is a stone ax). But recycling the beams didn't prevent a drastic deforestation of the area. By the time an extended drought hit in the twelfth century, the nearby building materials and pinyon nut sources had been greatly diminished, making the human suffering that much greater. By the end of the century, the residents had abandoned their fine city. We know only the bones of that story of disruption—it's just one of the cycles of human endeavor that have come and gone. But we would do well to pay attention, since we are in our own arc of the cycle of time, and we are making our own imprint upon that cycle: depleting, modifying, and multiplying the organisms and materials around us.

It's not possible to tease out all the details—or the hows and whys—of either human or juniper history. Both scientists and historians must take the limited evidence that exists and build a framework as best they can, working with the holes as well as the "facts," and covering the whole thing with flesh,

Spiraling on the Salt Flats.

like building a dinosaur model using a few bones for guidance. But the work of studying ecological history, though necessarily imperfect, has value: through it we can understand the deep history of the landscape better, and learn what this ongoing history might mean for the present and future.

Visitors to the Jardine Juniper stand before a current, tangible sign of a story that has been going on for more than a thousand years—one that rose out of millions of years of history. The strip of life curving up the dead trunk, the massive spiraling limbs, the shreds of bark, and little spray of foliage—they are precarious. No one knows how long the tree will live, but it is standing now, where past and future intersect. This is so for any tree, any landscape, any living being. The past spirals into the future in the present moment.

*

On our own visit to the tree, after we had sat in its presence a good long time, we finally left. On bikes, we flew downhill through the dark forest, past the spiraling grandmother tree, into a meadow of cheerful yellow flowers, scarlet gilia, penstemon, aspen. Choirs of crickets and tree songs. Blue melodic sky. The smell of a sunny meadow, the sound of wind coming up the canyon and the feel of it washing over you. Sage dancing, flowing in the wind. A whole field of waving, nodding mules ears.

We rolled down through shade and sun. Manzanita in bloom, the scent of its lacy flowers. Potentilla, yarrow, lupine. A family of grouse scattering, the young ones running with little peeps and the mother limping in the opposite direction with heart-rending cries. When she saw I wasn't following, she stopped limping and leisurely walked away.

5

Leaves and Seeds

In the 1990s I passed through Winslow, Arizona. Today, the town has taken advantage of the Eagles song "Take It Easy," with its famous line "standin' on a corner in Winslow, Arizona"; a statue of a Jackson Browne–type guy stands on a corner. Behind him, a trompe-l'oeil window reflects a girl in a flatbed Ford as she slows down to take a look.[1] But back then the town boosters hadn't touristed up the corner yet. Old brick buildings and a hodgepodge of businesses lined the historic Route 66, among them a pawn shop. I wandered into it and walked out holding a Navajo wedding basket.

I wondered about its history. Who had left it there? What straits had brought them to do it? I knew a little about the significance of the wedding basket, how it is woven, and the meaning of its design. I knew that wedding baskets are sacred in Navajo culture and are used in ceremonies throughout life. I wondered how this basket, with its spiraling symbolism of life's journey from birth to death, had nurtured the life cycles of a Navajo family. In the Navajo view, once a basket leaves Diné hands and its role in ceremony, it loses something essential. The weaver had never intended for it to become part of my own journey. But it had.

I have heard variously that the outer rim of a Navajo wedding basket represents protection, the entirety of life, the completion of life, a sealing of the journey, light. Some stories say that Changing Woman invented the herringbone-like weave of the outer rim. Changing Woman, a holy and powerful being to the Navajos, first appeared on earth just when the human race needed rejuvenation. She is the "embodiment of creation, the circular movement of

Navajo wedding basket, woven from sumac, with an edge resembling juniper leaves.

time," and she brings fertility, nurturing, and protection. She has the ability to be old or young, changing with the seasons.[2] According to one account, as Changing Woman sat under a juniper making the world's first wedding basket, she hesitated, not knowing how to finish it. Talking God came to her and reached into the tree. He broke off a sprig and handed it to her. She finished the basket to resemble the growth pattern of juniper leaves.[3]

*

Young Utah juniper plants (young as in ten to twenty years old, and perhaps one to two feet tall) bristle with short, needly leaves. They're sharp enough to bite the hand that grabs them. This prickly nature probably helps the young plants avoid being annihilated by deer and other browsers. Some juniper species never give up their spiky personalities, but when the junipers of the American West mature they clothe their branches with softer leaves—scalelike, tiny, and tapering. Utah juniper leaves lie close to the twig in opposite pairs, or sometimes in whorls of three. Each pair overlaps the next pair, making one of nature's elegant patterns, neat and precise—a lot like herringbone.

These leaves, of course, enclose the miracle of photosynthesis: within the tiny scales light, water, and carbon dioxide become sustenance and oxygen. The world is made more whole because of juniper leaves. For the gifts of photosynthesis alone we might stand in awe before any tree, or any blade of grass. How appropriate that the pattern of leaves—sustainers of life—rims the wedding basket.

The leaves of Utah juniper mature into a shade of yellowish green. Each tiny leaf has toothed edges, which you can't see with the naked eye. Under a microscope, though, they really do look like little teeth.[4] The leaves of the Utah juniper's close relatives also have "denticulate margins," but Rocky Mountain juniper leaves have smooth margins. This is an important detail—it's one of those characteristics that botanists used to decide that Rocky Mountain juniper is more closely related to eastern red cedar (*J. virginiana*) than to its sometime neighbor, Utah juniper.[5]

*

It took some years for me to appreciate and embrace juniper leaves and berries. Now they are indispensable to Christmas.

Long ago, every day after Halloween I would bundle up my little preschoolers and drive to Heber City to get there by 7:00 a.m. We'd get into line on the sidewalk behind guys in thick wool coats and stocking caps, women in big down parkas, and little kids muffled and gloved, waiting until 8:00 a.m. for the Forest Service office to open. I'd buy two permits to cut firs for Christmas, and then the day after Thanksgiving the family would head off on an epic snowy journey to find some perfect tree. We cut and decorated firs for a few years. Then the year came that I didn't have the energy to stand in that line. My dad had an alternative idea: we could get a pinyon permit instead.

The number of households in the West that have celebrated Christmas around pinyon pines must be considerable. Betty Wall, born in the 1920s, grew up in Lynndyl, a town in Utah's West Desert. Lynndyl is still tiny, with a population of 103 as of 2010. When Betty was a girl, in early winter young men went up into the hills with a team and wagon to chop Christmas pinyons. They'd come back with a load piled high. Since Betty's family lived closest to

The overlapping, scalelike leaves of Utah juniper.

the mountain, her mother could flag the guys down and get first pick, paying them about thirty-five cents for a tree that filled the house with the smell of forest. "We used to put wax candles on the tree. It's a wonder we didn't burn all the houses down," Betty says.

Seventy years later, we were going to have to get our own pinyon by driving south out of Salt Lake past Provo and then veering east up Spanish Fork Canyon on its winding, endless two-laner. Then down through the Castle Gate rock formation, where railroad cars still load up with coal, through Price and out through Wellington, where we'd turn north and head for the then-dirt road that travels through Nine Mile Canyon, famous now as "the world's longest art gallery," with its thousands of prehistoric rock art images. Though oil and gas trucks spin along the road now, in the early 1990s the canyon was a quiet place: a few ranch houses here and there, alfalfa fields and pastures along the valley, and mountains and cliffs on either side.

In the thirteenth century, the Florentine artist Cimabue painted for the glory of God and left behind paintings, altarpieces, frescoes, and crucifixes now preserved in churches and museums. Working in that same century and before, members of the Fremont culture of the Colorado Plateau made art for we know not what. But they left behind their work, and we can at least stand before their images and imagine what they meant to say. Which we did in Nine Mile Canyon, the four of us: my dad, my sons Jed and Will, and I. Then we sat at a picnic table and ate turkey sandwiches and soup from a thermos.

After lunch, the sky started to dribble snow, and it was time to get some trees and get out of there. The woman at the BLM office in Price had given us a map of an area where we could cut, up a side canyon. We bounced up this spiny side road for a while but before long realized we'd better stop pushing our luck and get out and walk. The boys carried the bow saws and we began to hunt, climbing the steep canyon sides through sagebrush and rabbitbrush, snow patches and cold damp soil, and of course through cedar trees everywhere. We wandered and slipped in the mud all over that canyon, and the snow kept spattering down.

Not a pinyon in sight.

It occurred to us that the woman in the BLM office had said we could cut either a pinyon or a juniper. But really—a juniper? Homely, ungainly, not-really-the-right-kind-of-conifer tree? Juniper is not exactly the Little Fir Tree or anything that had anything to do with memory, tradition, or homey Christmas mornings with all the family gathered. Its leaves and shape were all wrong. But the dark of late afternoon was smudging the canyon air. And we couldn't leave empty handed. So—we shifted the paradigm and began looking at the cedars: the plump exuberant ones, the twisted old ones, the ungainly lopsided "young" ones. We took note of the pale blue berries, the sculptural forms, the fragrance of the tiny leaf-clad twigs, and thought maybe one of these would do.

My father found a tall Eiffel-tower type. For our house we cut a shorter one—asymmetrical but the best we could find. We got them tied on the roof of the Blazer and headed home, wondering what the people at home would say.

I will skip right to their delight at the wild, imperfect desert presence those trees brought into both houses. And how after that, nobody wanted anything but juniper and asymmetry for Christmas. And how the day after Thanksgiving forever after became a pilgrimage to Utah's West Desert with turkey sandwiches, hot soup, and leftover pumpkin pie; driving rutted snowy roads; wandering through gnarled woodlands with tree seeking as an excuse to be together in the open sky and the spare November day; cutting into the pale fragrant wood; tying ungainly stiff branches to a car top; then looking out across the silent landscape together and wanting to keep that specific moment always.

A group and their Christmas tree, cut in the Onaqui Mountains.

If you yourself cut a juniper for Christmas, in-laws and neighbors may think you odd, but actually you would not be completely alone. Even during the 1950s, that very conventional decade, a number of unconventional—or perhaps thrifty—folks sought out juniper trees. In 1959, the Tooele County, Utah, stringer to the *Deseret News* promoted the tree (and his home county) by writing, "Comes Christmas time and Utahns are sure to be lured to seek pitch-incensed juniper . . . [through] a convenient, enhancing short drive by automobile from our cities and towns to nearby or neighboring juniper-studded foothills of the . . . Onaquis, Stansburys, Simpson Buttes, Ibapahs, and far from least, the famed Cedar Mountains."

Wandering the West Desert in search of a tree, the author promised, was "a wonderfully enjoyable way to spend a day of recreation—real fun—to motor out to the junipers—out away from the hustle-bustle urban routine." One family living on a ranch near Pilot Peak cut their "juniper Tannenbaums" on that mountain; another family made an annual trip to their "favorite juniper valley" at Johnson Pass, between the Stansbury and Onaqui Mountains, to gather "several berry-speckled, lacy-foliaged juniper branches placed as decorations around the house. To them it brings the incensed, green outdoors right into the cozy interior of the family's favorite nooks and dens."

After reading that last sentence, how could one desire anything but a juniper? The writer ended by saying, "No other tree, it seems, is so well fitted as this one to endure the arid, wind-blown, sand-swept land of Deseret. Certainly, none is so highly incensed that it brings its alluring fragrance right inside the house and onto the mantle as does the juniper."[6]

*

Unconventional, yes, but evergreen juniper leaves do what all Christmas greenery is supposed to do—symbolize life during the dead cold of the winter solstice. The custom of Christmas trees originated in an era when nature held more psychological power than it does today; in long-ago times, Scandinavian peoples brought evergreen boughs and trees into their homes for protection. On the American continent, ancient people brought juniper seeds and leaves into their houses for many reasons, including, perhaps, protection. At Salmon Ruin in New Mexico, first occupied almost one thousand years ago, archaeologists gathered and studied more than 1,600 Utah juniper cones and seeds from the ruin. They found these in trash pits and burials, mingled with debris from roof falls, and in storerooms. Some had been charred, some had not. Some had insect-bored holes in them. Rodents could have brought some of them into the ruin, but archaeologists believed most of the seeds were part of a food supply, since they were found near other foods. In one kiva, cones and seeds were found near maize and grinding stones. The edible parts of juniper cones, according to these researchers, could provide around 5,500 calories (kcal) per kilogram of seeds. Therefore, based on the woodlands in the area, the junipers could have provided half the calories required by the three hundred people living in this community.[7]

The people also used the seeds in burials, as adornment, or in ceremonies. At Mesa Verde, archaeologists have found juniper cones beneath the clay floors of some rooms. This is reminiscent of a Hopi practice of spreading meal and piki bread crumbs (made from cornmeal and ash from juniper twigs or cones) on the ground before laying a floor in a house. At Salmon Ruin, the people buried their loved ones wrapped, along with juniper seeds, in juniper-bark matting. In addition, someone left behind in the ruins a neatly wrapped bundle of juniper twigs—perhaps prepared for medicinal or ceremonial use.[8]

*

In 1653 in England, Nicholas Culpeper published a comprehensive guide to herbs. He had nothing but good to say about juniper's qualities:

> This admirable solar shrub is scarce to be paralleled for its virtues. The berries are hot in the third degree, and dry but in the first, being a most admirable counter poison, and as great a resister of the pestilence, as any growing: they are excellent good against the biting of venomous beasts: they provoke urine exceedingly, and therefore are very available to dysuries and stranguries. It is so powerful a remedy against the dropsy, that the very lee made of the ashes of the herb being drank, cures the disease: it provokes the terms, helps the fits of the mother, strengthens the stomach exceedingly, and expels the wind; indeed there is scarce a better remedy for wind in any part of the body, or the cholic, than the chemical oil drawn from the berries. Such country people as know not how to draw the chemical oil, may content themselves by eating ten or a dozen of the ripe berries every morning fasting. They are admirable good for a cough, shortness of breath, and consumptions, pains in the belly, ruptures, cramps, and convulsions. They give safe and speedy delivery to women with child: they strengthen the brain exceedingly, help the memory, and fortify the sight by strengthening the optic nerves: are excellent good in all sorts of agues, help the gout and sciatica, and strengthen the limbs of the body. The ashes of the wood is a speedy remedy to such as have the scurvy, to rub their gums with. The berries stay all fluxes, help the hæmorrhoids or piles, and kill worms in children. A lee made of the ashes of the wood, and the body bathed therewith, cures the itch, scabs, and leprosy. The berries break the stone, procure appetite when it is lost, and are excellent good for all palsies, and falling sickness.[9]

All of those virtues in one homely plant sound fantastical. But then again, what do *we* know? We who go about our indoor lives of working, eating, and sleeping; we who pass through nature occasionally, remarking on beauty or thrilling ourselves with speed and bumps provided by gravity, wind, or fossil fuel; who get a little fresh air and exercise and then go back inside and take a pill or vitamin or drink for whatever ails us. Culpeper, on the other hand, spent his life walking the countryside, learning to know plants firsthand and intimately, and treating thousands of patients with these plants. So we may at least be open to the possibility of juniper's medicinal qualities.

One clue that juniper may indeed "be scarce paralleled for its virtues" is the fact that the indigenous peoples of the western Americas used their own local juniper species in many of the same ways that Culpeper recommended. A search of even one species of juniper—*Juniperus osteosperma*—in the huge Native American Ethnobotany Database maintained by the University of Michigan–Dearborn brings up eighty-seven matches. Below are some of the data. In no way, however, should this list—or Culpeper's—be considered medical or culinary advice. This is a report of ethnobotany, not medicine. Some species of juniper are toxic and should *not* be used. So, as with all herbs, you should know exactly what you're doing before ingesting juniper in any form.[10] With that explicit caution, here are juniper uses from the ethnobotany database: The Havasupai used a tea of the green twigs to treat colds; the Hopi used a similar tea for women in childbirth.

Navajos ate the seeds for headaches and used the tea for washing hair.

Paiutes used tea from young twigs for menstrual cramps, hemorrhages, and stomachache, and they drank tea or breathed fumes from burning twigs for headaches and colds. Berry or twig tea eased coughs, fevers, influenza, and pneumonia. Paiutes used the branches in a sweat bath for colds and rheumatism. They also used a decoction of the berries as a wash or in poultices for their rheumatic aches. A poultice of mashed twigs could ease swellings or rheumatism, draw out boils and slivers, and treat burns. Steam from roasted berries was supposed to ease pains. A strong decoction could provide an antiseptic wash. A tea from the berries acted as a diuretic. A decoction of twigs helped with kidney trouble.

The Shoshones used it similarly: they drank tea from berries or twigs as a blood tonic and to treat venereal disease and worms. They also used this decoction to wash the body during measles and smallpox. They used it for heart and kidney troubles. A poultice of leaves could be held to the jaw for swollen and sore gums and toothache, or to the neck for a sore throat.[11]

A modern-day witness, Dorena Martineau, cultural specialist for the Paiute Indian Tribe of Utah, told me that juniper "is a very worthy tree and is still used to this day by many American Indians." People still commonly use the leaves or seeds for colds and coughs. They use berry "broth" to treat asthma, act as a diuretic, or help indigestion. The juniper berries, she maintains, taste sweet, not bitter.[12] And maybe this is true, though it's a piney sort of sweetness.

In short, the leaves and seeds of western species of juniper have been used among various western peoples as an emetic, as a contraceptive, and in sweat baths, and for ailments as diverse as stomachache, constipation and diarrhea, worms, spider bites, burns, measles, boils, hives, sores, insects in the ear, tooth decay (tree gum was used as a filling), heart trouble, fevers, malaria, pneumonia, tuberculosis, epidemics, venereal disease, diabetes, menstrual cramps and bleeding, childbirth pain, postpartum issues, bruises, sprains, backache, headache, rheumatism, swelling, sores, fainting, and lack of energy.[13]

Folklorist Barre Toelken reminds us that the native people probably saw and used these medicines differently and more holistically than we do. "We [Anglos] may have a few home remedies, but for most big things we go to a doctor. A Navajo goes to the equivalent of a priest to get well because one needs not only medicine, the Navajo would say, but one needs to reestablish his relationship with the rhythms of nature. It is the ritual as well as the medicine which gets one back 'in shape.' The medicine may cure the symptoms, but it won't cure you. It does not put you back in step with the things, back in the natural cycles—this is a job for the singer," that is, the medicine man.[14]

*

Besides medicine, juniper provides food. Acoma Puebloans cooked the berries into a stew; Goshutes boiled them and ate them in the fall and winter; Havasupai dried and stored the berries and used them to make a drink. Hopis ate them with piki bread. The Tubatulabal of California used them "extensively" for food. The Yavapai ground the berries and made them into a meal, added water, and drank the beverage, or they made the meal into cakes.[15]

In June 1776, Padre Francisco Tomás Hermenegildo Garcés saw firsthand the use of juniper for food in the Southwest. This devout priest, born in Spain and inclined to God from his earliest years, had become known as the "Children's Priest" for his "simple and artless" nature. At age twenty-four he had volunteered to come to New Spain to convert American Indians.[16] He is said to have loved the natives he came to serve, but he also loved to explore, it seems. In 1776 he journeyed from the San Xavier del Bac Mission (near present-day Tucson) to visit the Hopi mesas. On his way, he traveled through land "clothed with junipers and pines" to Havasu Canyon and visited with the Havasupai people.

He stayed for five days at the beautiful spring in the canyon. During this time, I imagine that he gazed at the turquoise creek and pools, marveled at the waterfalls plunging down cliffs, and felt rest in this place, or even the holiness he so tirelessly sought. The "Jabesúas" treated him with great kindness. They "waited upon me and regaled me with flesh of deer and of cow, with maize, beans, quelites [greens, perhaps], and mezcal [distilled from the agave plant], with all of which they were well provided." Though the people had access to all these fine foods, "they also eat a berry of the juniper, a tree which is very abundant in these lands."

The climb out of Havasu Canyon, on a trail presumably much narrower than today's precipitous path, gave him some moments of "horror," but then he traveled "over good ground, with much grass, and many junipers, pines, and other trees" to a "rancheria whither had come some of this nation to gather the fruit of the juniper." If the cliffs of Havasu gave him a fright, he had worse ahead. After making several substantial forays into the "wilderness" of the Southwest, his mission ended when Yumas, rising up against Spanish oppression, clubbed him to death in 1781.[17]

Shirlee Silversmith, director of the Utah Division of Indian Affairs, told me that she, like many other American Indians, always keeps a jar of juniper ash in her kitchen to use in several traditional dishes. Traditional corn bread and corn mush particularly need the alkali processing that juniper ash provides. Juniper ash water helps release the hull from the corn, adds minerals (calcium, potassium, phosphorus, iron, zinc, and many more), makes the corn protein more usable, and increases the absorbability of niacin.[18] But using juniper ash is about more than nutrition. Silversmith told me that in the traditional view, everything is connected to us, and what we take into our body benefits us because it strengthens that relationship. "It's a matter of respect," she said. "When kids are taught how to gather herbs, they offer prayers and promise to replant. We don't go out to destroy. On the reservation you build homes around plants and rocks; you don't have to dig and tear down trees. That's traditional living."[19]

One recipe for making corn bread with juniper ash water goes like this: Gather branchlets of the juniper. Set the needles of the branches on fire, holding them over something like a skillet to catch the ashes. Burn only the needles. When you have a cup of ashes, add one cup of boiling water to them. Stir, and strain out the ashes. Add the water to 3½ cups of boiling water, then add six cups of blue cornmeal. Knead the dough, shape it into flat loaves, set these loaves onto the hot ashes of a fire, and bake for one hour. (It's okay to use a skillet if you don't have a fire.) To make paper-thin bread, use more water. To make mush, just add a few handfuls of cornmeal to the water and stir and cook until thick.[20]

Besides the food they could directly provide, pinyon-juniper, or PJ, woodlands also harbored more plant food resources for indigenous peoples than other ecosystems in the interior West. In the pinyon-juniper woodlands on Black Mesa in northern Arizona, investigators found numerous birds and animals and twelve different plant foods, including pinyon nuts (of course), banana yucca, Gambel oak, ricegrass, Mormon tea, and prickly pear.[21]

*

If you crush juniper berries or leaves, they will release a fresh conifer scent because of the turpenes they contain. Taste them, and they release their aromatic flavor into your sinuses. It is not a taste that is available in fast food, or even most slow food, but chefs do use juniper berries to season wild game dishes.

In turn, wild game use juniper, even though the aromatic turpenes make it a less-than-ideal food. Mule deer depend a lot on PJ. When snow and cold make life tough and food scarce, the deer inevitably turn to juniper. Anywhere that deer overwinter, you can see how they have "pruned" the branches of the

A juniper in the foothills above Salt Lake City, "pruned" of its lower branches by deer; nearby a young juniper with sharp leaves is untouched.

cedars, giving the foliage a nice flat bottom, as high as the deer can stretch their necks. Utah juniper does have nutritional value for the deer, but it's not the most nourishing plant in the world. It is high in calcium but low in phosphorus. It has 6.4 to 7.9 percent protein—and deer need 8 percent just for maintenance. Finally, the volatile oils can harm the microorganisms that help the deer digest their food. Still, sometimes juniper is the only choice deer have, and it can mean the difference between making it to spring or dying of starvation.[22]

Juniper and pinyon-juniper woodlands also keep the deer alive by giving them critical cover and shelter during the winter. As the animals huddle in the trees, somewhat shielded from winds and blizzards, they can conserve their hard-gotten calories. Because deer rely so much on PJ woodlands, the people responsible for managing the deer herds (for the benefit of hunters, largely) have an interest in what happens to these ecosystems. Forty years ago, biologists undertook a study to find out the effects of juniper removal on mule deer.

Removal projects had become common in Utah, mostly for the purpose of increasing rangeland for cattle. The biologists looked at widely separated sites in Utah: the Sanpete Valley, Book Cliffs, and Natural Bridges areas. In each of these places they studied fecal pellet groups in both cleared sites and natural PJ woodlands. Here is a pursuit you can undertake on your next hike: count fecal pellet groups—or PGs, for short. One PG is ten droppings or more of the same size, shape, and color. You will note a difference between winter and spring PGs: in winter the pellets are hard and cylindrical, reflecting the toughness of winter browse; the more tender plants of spring and summer make for softer, misshapen pellets.[23]

The investigators also sat with spotting scopes and binoculars and observed deer numbers, locations, and activities, and they counted tracks. They watched each deer for five minutes and recorded how long it fed on which plants. They did this cold work for three winters.

In the end, they found different results in each area. In Sanpete Valley, located in central Utah, the deer used the natural PJ woodland much more than they used the cleared areas, especially during the daytime. At Natural Bridges, at the southern edge of the state, there was no difference in deer use between the treated and untreated areas. And in the Book Cliffs, a remote high area west of the Green River, the deer used the treated areas more than they used the natural areas. In other words, generalizations don't work.

Terrain and weather likely make the difference. The deer tended to avoid large, flat areas, preferring protected areas with valleys and swales where they could gather and find cover and a sense of security.[24] The animals gathered on southern slopes that had been cleared of trees so that they could catch more sunlight, but they retreated to the woodlands when snow was deep, temperatures dropped below twenty degrees Fahrenheit, or winds raged. Their instinct, it seems, was to try to minimize heat loss rather than take in more calories. As for food, they did eat a great deal of juniper—in some cases, they preferred it—and also sagebrush, Russian thistle, mountain mahogany, and grasses, when they could get them. They browsed and bedded down in about two-hour intervals. And thus, they spent their winter lives among the junipers, wandering not very far, trying to stay warm and fed through the storms and cold.[25]

*

Across the ocean, another animal is loving juniper leaves and berries to death. While rabbits in the West help junipers spread by eating and pooping seeds, in the United Kingdom too many rabbits are devouring juniper shrubs, decimating the juniper population in some areas. Some people think that the rabbits are even endangering the production of gin, which requires juniper berries for its distinctive flavor. And in fact the name "gin" derives from a foreign word for juniper, either French (*genièvre*), Dutch (*jenever*), or Italian (*ginepro*). Invented some four hundred years ago as a medicine, the drink was strong and cheap, and it quickly gained a following among the lower classes. The potent taste of juniper berries somewhat disguised the fact that the alcohol used in gin was low budget and harsh.

We'll never know how large a role juniper berries played in helping people get through Prohibition in the United States, but it would be fair to surmise that juniper flavored countless gallons of "bathtub gin." I visualize my spunky great-grandmother Turner picking pails of berries and using them in the spirits she distilled in the cellar of her adobe house. She lived by the railroad tracks in the Mormon village of American Fork, Utah, and her grandchildren were not allowed to visit her on their own. The younger ones never knew why until much later, when an older cousin said, "Oh! Didn't you know about Grandma's still?"

Since then, gin has turned into a sizable industry. And since in England rabbits are eating an essential ingredient for gin and tonic—"Britain's favourite drink"—the alarmed makers of No. 3 London Dry Gin have donated £1,000 for rabbit-proof fences around dwindling juniper stands in West Sussex. The U.K. charity Plantlife is also working to protect the British junipers. Junipers were one of the first trees to colonize the island after the last major Ice Age, says Tim Wilkins of Plantlife, but during the last few decades the trees have steadily declined in Britain. Within fifty years junipers may disappear from much of England. "Such a calamity would represent more than the loss of a single plant type—it supports more than 40 species of insects and fungus that cannot survive without it."[26] You can't blame the rabbits alone, though. A number of strands of this ecosystem web have been tugged. When fewer animals graze an area, the grasses get denser, and juniper seedlings can't compete with the grasses. Also, a fatal fungus has been attacking junipers; *Phytophthora austrocedrae* infects the roots and destroys the foliage. As a result of many factors, from 27 to 58 percent of juniper has been lost in the various countries of the British Isles during the last sixteen years.[27]

Organisms that we take for granted on the landscape are not as permanent as we imagine. Any number of disruptions can change things. And we may not even notice the disruptions and changes—unless they affect our own profits or pleasures.

*

The ecosystem web of western North America differs from that in Britain, and nobody here minds if animals eat the junipers. But any one action affects the whole, and there's no predicting exactly how. The intrepid, winter-enduring authors of the Utah deer study noted that of the 744 vertebrates living in Utah (371 birds, 247 mammals, 59 reptiles, 55 fish, and 12 amphibians), the majority live in or are influenced by the pinyon-juniper ecosystem. So far, they said, research had touched on only a dozen species of PJ-influenced mammals, with inconclusive results. So no one can understand the full consequences of manipulating pinyon-juniper woodlands. The authors concluded by saying: "When we consider . . . the several hundred animal species that are distributed over an array of p-j subtypes, with each subtype having peculiar post-use histories, and literally dozens of potential treatment methods; what can we say but that we sit in the most extreme state of ignorance?"[28]

6

Wood

April 2011, driving north on Highway 89 in the twilight: all along the highway, cedar post and barbed-wire fences enclose big swaths of land, keeping things in, keeping things out. Sometime decades ago, a rancher cut the trees down, cut the lengths, dug the holes, set the poles, stretched the wire. Rain clouds now hang over the Paunsaugunt Plateau. Dark cedars on dark pink hills line the valley. The land at this time of day is mysterious and wide. Naked cottonwoods stand ghostly along the Sevier River. Misty cliffs to the north. Sagebrush. A ranch here and there. Signs for development property or lots every now and then. Someday, I think, I may look back on this very moment of flying along Highway 89, now so empty and shadowy in the twilight, and wonder how a place could change so much. There may come a time when we will miss the openness and the old pale and crooked cedar poles.

The cedar post fenced the West. It contained the West and changed the landscape. It became the visible—and enduring—evidence of changing human endeavors on the land. Fences manifested the fact that inhabitants were losing their freedom to roam. Agriculture and the civilization made possible by agriculture would inexorably replace gathering and hunting lifeways.

*

The wood of the cedar post is a marvel in itself. Leonard Read writes that we can't even make a tree, let alone describe one, "except in superficial terms. We can say, for instance, that a certain molecular configuration manifests itself as a tree. But what mind is there among men that could even record, let alone

Cedar (juniper) post and barbed-wire fence in Utah's Rush Valley.

direct, the constant changes in molecules that transpire in the life span of a tree? Such a feat is utterly unthinkable!"[1]

And that's true, but sometimes we need those superficial descriptions. Experts describe the Utah juniper, the tree that gave the developing West most of its fence posts, in this kind of language:

- A small monoecious (sometimes dioecious) tree or arborescent shrub, 3–6 m tall . . . sometimes several upright branches, nearly the same size as the main stem, arising from near ground level, forming rounded clumps or crowns (5–8 branches at 1/5 m height), branchlets stiff, twigs relatively stout . . . bark reddish brown or gray brown, weathering ashy-white, thin, fibrous, and shreddy in long strips.[2]
- A short tree that may live as long as 650 years. Utah junipers grow less than 26.4 feet (8 m) and are often as short as 9.9 to 14.85 feet (3–4.5 m), with a trunk 4 to 7.5 inches (10–30 cm) thick. Sometimes the tree has multiple stems.[3]

Close-up of old, weathered juniper wood with bark stripped away.

- Its branchlets are stiff and stout, and its short, often forked trunks are covered with thick red-brown bark that sloughs off in shaggy, fibrous strips. . . . These spreading, round-crowned trees are seldom over thirty feet in height and thirty inches in diameter, as most of the larger trees have long since been harvested. . . . The sapwood is almost white in color, but the aromatic heartwood contains an accumulation of chemical substances that color it deep red. Thus Utah juniper—and other junipers as well—are often called cedars.[4]

You can't grasp the essence of a tree with any amount of precise language, observation, or measurement. But then, neither can you capture its essence by only sitting beneath it and meditating on nature. Or by building fences with it. Observational, intuitive, artistic, and practical approaches all contribute to our understanding of the world. It's easy to become proficient in one or two of these approaches. It's not so easy to step out of habit and engage with the world in a different mode. Henry David Thoreau, whose mystical love of nature could have completely consumed him, walked this line of balance. He knew trees not only by walking in rapture through forests, but also by working wood with his hands and by observing the environment with a naturalist's eye. The naturalist in Thoreau advised that we should "not underrate the value of a fact; it will one day flower in a truth."[5] So therefore, on with some facts about wood.

Wood is, of course, the aspect of a tree that gives it shape and structure. As the framework of the tree, wood must perform specific functions to enable the tree to thrive, and in order to perform these functions it must have an exceptional and rare combination of qualities: its strength helps the tree survive high winds; its stiffness keeps the tree upright without sagging; its lightness allows the tree to grow tall without collapsing in on itself. Wood endures blows and other damage without shattering. Its unique cellular structure allows it to bear the great weight of lateral branches and the bending force of wind. Compare brick, plastic, glass, steel, fiberglass, concrete. None of these can do all of this. The website for the Natural History Museum of Britain, in fact, once gushed that "weight for weight, wood has probably the best engineering properties of any material."[6] This is not actually true; materials like carbon or Kevlar composites and several kinds of metals can outperform wood in terms of lightness, strength, load bearing, and flexibility. But of course if you need a light, strong, flexible substance that can also transport nutrients and heal damage, only the wood of a living tree will do.

The macrostructure of this material called wood is the same in every tree, with these layers from the bark inward:

Bark. This provides a protective outer cover.

Phloem or inner bark. This layer carries the food produced by the leaves or needles to the other parts of the tree.

Cambium. This layer produces new phloem and xylem cells and so is responsible for the tree's growth in girth.

Xylem or sapwood. Through this layer water flows from the roots upward. This layer also provides strength.

Heartwood. The site of inactive and dead cells, this layer is darker in color because of mineral deposits, gums, and resins. Heartwood also gives strength to the tree.

Pith. The remains of the tiny sapling from which the tree originally grew form the material at the center of the trunk.

*

Humans live in and by the material world, using and fashioning all sorts of materials for our purposes, from toothpicks to space stations. The first materials we used for our purposes were rocks, soil, plants, and animals. In all our producing and consuming of today, with all the substances we have invented and use for our needs, wants, and non-wants, we tend to overlook our primal relationship, indeed our utter dependence, on these basic materials. The word "material" means "real," "ordinary," and "earthly," or "substance, matter from which a thing is made." Etymologically, the word "material" derives from the Latin *materia. Materia* originally referred to the hard inner wood of a tree, the essential material for making many things. The Latin *materia,* in turn, is kin to or derives from the word *mater*—"origin, source, mother"—and *matrix*—"womb, source, origin." In our very language and endeavors, we have echoes of the "mother tree."[7]

This wood, this *material,* has sheltered people and animals for as long as they have lived on earth. The Salmon and Aztec Ruins of northwestern New Mexico remain as examples of one era—around AD 1100—when people used thousands of junipers and other trees to build multilevel homes. The builders in these communities used juniper or ponderosa logs as beams—or vigas—for ceilings and floors. To make a ceiling (which might in turn be a floor for an upper story), they inserted the vigas into sockets on the walls and then laid smaller branches—or latillas—on top of the logs at right angles. On top of the latillas, they would lay willow or other twigs close together, again at right angles, and cover the whole thing with a thick layer of adobe. They would then put down a layer of shredded juniper bark, and then another layer of adobe.[8]

Ancient Puebloans also used pinyon, cottonwood, Douglas-fir, and other trees in their structures. For the great houses of the Chacoan period (built between AD 1000 and 1100), both at Chaco and in outlying communities like Salmon and Aztec, the people had to go miles to get trees large enough to span large spaces. For smaller rooms, they could use the materials close at hand—and they used mainly juniper. Although pinyon and cottonwood might also

Juniper vigas overlaid with latillas in ceiling of an early-ninth-century cliff dwelling in the Cedar Mesa area. The ceiling is original and probably looks very similar to how it looked 1,200 years ago. Courtesy Bureau of Land Management, Utah.

have grown nearby, juniper suited their needs best. Pinyon grows too crookedly to be used for vigas, but a typical juniper could provide one viga or a latilla or two for a moderate-sized room. One room might use 6 to 8 vigas and 26 to 50 latillas. (Imagine how long it would take to cut the trees for even one room!) The builders could use smaller branches to construct lintels over doors and smoke vents; with the thick walls of pueblo construction, they would need 8 to 14 branches for each opening. They might use more branches as small splints above the latillas. At times, a room might also require posts, and junipers made good, long-lasting posts. In thousand-year-old ruins, you can still see original beams and posts.[9]

Incredibly, researchers estimate that for the original eleventh-century construction at Salmon—three stories, with some three hundred rooms as well as kivas—the builders would have needed 7,500 to 9,500 trees, mostly local

Juniper logs used in construction of Ancestral Puebloan structure, Owl Canyon, Utah.

juniper but also ponderosa brought in from miles away. After about AD 1120, later occupants extensively remodeled the original construction. They reused a lot of the original wood, but they would have needed 2,900 to 5,600 new trees as well. For whatever reason, these later people didn't go after the distant ponderosa logs but relied mainly on the nearby junipers.[10]

The impact on the landscape of all these trees cut, along with those needed for firewood, must have been tremendous. As people sought to live comfortable or even subsistence-level lives, the woodlands would have shrunk, year by year. The Puebloans depended on the trees, and at the same time they had to cut them. Did they see what the inevitable depletion would eventually do to their culture? Did they try to do something about the coming future or did they just use what they needed without worrying?

As the forests shrank, the neighboring villages would have had to compete for the same wood resources. The villages may have argued or fought over "lumbering rights." Too many people competing for too few resources—these are the same fights that have shaped human history to the present moment.

*

No doubt the cutting of trees exacerbated the miseries that apparently came upon the people of the Southwest during the thirteenth century. In addition to the resource depletion, the climate turned unpredictable during that century—the worst event being some thirty years of severe drought. Imagine being so hungry that you would eat juniper bark. The inhabitants of Salmon who tried to hang on during the drought did that very thing; the feces (called coprolites) they left behind for archaeologists to pick apart and examine contained partially roasted juniper twigs, juniper bark, yucca leaves, maize cobs, bones, and insects. Juniper bark and twigs don't provide a lot of nutrients, but apparently they could help ease an aching hunger.[11]

In happier times, juniper bark, especially the bark of old junipers, had to serve only day-to-day, nonedible uses. Two thousand years ago, Basketmaker II people made unfired pots by strengthening the clay with juniper bark. In the Salmon Ruin, people shredded bark to use in a number of ways, including the making of cordage, bags, mats, and sandals, and for pot rests and toilet paper. It padded babies' bottoms, too. In these ruins, archaeologists found masses of shredded bark in several rooms, stored for future use. These people used *materia* for many of the same basic needs that we have.[12]

For the Havasupai, writes Donald Peattie, juniper bark *materia* took people from birth to death. When a boy was born, his mother covered him with a blanket made from juniper bark rubbed very soft. He would sleep on a mat woven from bark, and the winter fire that warmed him would be made using a tinder of dried bark. At puberty he would run toward the dawn carrying a slow-match made of twisted bark; he would touch his ankles, knees, wrists, and elbows with the fire. He would bring his mate to a marriage bed of juniper bark.[13] In the end, an individual might be buried wrapped in a mat woven from juniper bark fiber. Farther north, on the southwest shore of Utah Lake, a middle-aged man who died around 4,700 years ago was buried with an open-twined juniper mat. With obvious respect and care, those he left behind laid him to rest also with antler and bone tools, projectile points, a basket, and a dog.[14]

Plate 1. Twisted juniper overlooking Dead Horse Point.

Plate 2. The skeletons of Utah junipers on a slope in the Great Basin.

Plate 3. A juniper that has found a roothold amid hoodoos near Bryce Canyon, Utah.

Plate 4. Roots that have worked their way between rocks.

Plate 5. An ancient, grandmotherly juniper at Arches National Monument.

Plate 6: The trunk of the grandmotherly tree, with its streaked wood, ragged bark, and desert sand lodged in its folds.

Plate 7. Each juniper grows into a unique form over decades and centuries, nudged by unique circumstances and events.

Plate 8. A tree squeezes through a crack where a seed lodged centuries ago in Needles District, Canyonlands National Park.

Plate 9: Curve of the wood grain in a standing juniper.

Plate 10: Color variations in the wood grain of a juniper.

Plate 11. Subtle forms and colors in the wood grain of a juniper.

Plate 12. At left, a pinyon pine growing in association with juniper, as is common throughout the interior West.

Plate 13. Mature juniper berry beside a small cactus and dried pinyon pine needles.

Plate 14. Juniper plant community in Arches National Park.

Plate 15. Spiky leaves on the twigs of a young juniper.

Plate 16. Utah juniper berry and twigs, showing the herringbone pattern of the overlapping leaves.

Plate 17. Juniper berries (botanically, cones) and leaves, with brown male cones at the ends of the twiglets.

Plate 18. Dried juniper twigs, seeds, and "ghost beads"—seeds with a hole in one end—found at the base of an old tree.

Plate 19. Rocky Mountain juniper.

Plate 20. In the background, juniper woodlands darken the lower slopes of the Wasatch Plateau.

Plate 21. Thick juniper cover on Cedar Mesa, Utah.

Plate 22. A scattering of junipers outside Capitol Reef National Park.

Plate 23. Juniper tree with the iconic Delicate Arch in the background.

Plate 24. The growth and architecture of an organism made visible:
a long-standing dead juniper at Dead Horse Point.

*

We know about drought and the ages of ancient structures partly through the science of dendrochronology, or analysis of tree rings. To state it perhaps too simply, dendrochronologists can compare tree-ring patterns in trees whose lives overlapped, going back through time perhaps a millennium or two, and then cross-date those trees to pinpoint their ages. In arid regions, they can compare the rings of trees that recently lived to historical rainfall records. Putting all the information together, they can reconstruct centuries of climate information.

The rings are formed as cells divide, as the tree grows. In conifers, long, thin cells called tracheids form the wood. The tracheids are hollow, with a matrix of tiny cellulose fibers strengthening the cell walls—a combination that makes for stiffness, lightness, and strength. The cells that form during the springtime, with its more abundant moisture, have thin walls and larger interiors. The cells formed later in the season are smaller, with thicker walls. Because of these thicker cell walls and the fact that late-season cells are more packed together, this "late wood" looks darker than early wood. In years with plenty of moisture, more cells divide, making rings thicker than those formed during drought years.[15]

"It is almost a marvel that trees should live to become the oldest of living things," wrote Enos Mills in 1909. "Fastened in one place, their struggle is incessant and severe. From the moment a baby tree is born—from the instant it casts its tiny shadow upon the ground—until death, it is in danger from insects and animals. It cannot move to avoid danger. It cannot run away to escape enemies. Fixed in one spot, almost helpless, it must endure flood and drought, fire and storm, insects and earthquakes, or die." If a tree does survive onslaughts of nature and humans, it does so largely through the resilient qualities of wood. Juniper in particular has compounds (hexane soluble and methanol soluble) that make it fragrant and help it resist rot and insects.[16]

When we cut our Christmas juniper, I am often amazed—and feel a twinge of remorse—at the number of rings we can count in these modest-looking trees. Although in good soil and precipitation a Utah juniper can grow more quickly, the wood might grow outward at 0.05 inch in diameter per year, or even more slowly. A twenty-year-old tree, then, may have a trunk less than a couple of inches thick.[17]

The shreddy bark from older junipers has been used in numerous ways over the centuries.

The ridged bark of an old juniper tree.

Juniper bark that has weathered in a checked pattern.

In a pinyon-juniper woodland in Mesa Verde, researchers dated a 32-inch pinyon pine at 431 years old. On the other hand, they determined that a 15-inch-diameter juniper was 400 years old—in other words, almost as old as a pinyon twice its size. Since the largest juniper in that stand had a 36-inch diameter, it appears that the oldest junipers in the area are several hundred years older than the oldest pinyons. Why? Were junipers more able to survive drought, porcupines, bark beetles, root fungus, fire, or blight? The researchers could not answer, but it looks as though juniper had a survival advantage in that environment.[18]

But no one knows for sure how old a Utah juniper can grow to be. In old age, these trees are difficult or impossible to date. For one thing, they may begin branching right at ground level, or they may grow erratically. The trunks may not be round but lobed. Where the climate is especially dry, during many years the tree may not even produce a growth ring. On the other hand, two wet seasons in one year could stimulate the tree to create two growth rings. Finally, when a tree is very young, it may not produce growth rings at all as it waits for the right conditions to grow.

If the oldest Utah junipers could be dated accurately, we might find that this species can live a very long time. Its cousins western juniper and Rocky Mountain juniper, which grow more symmetrically, can be accurately dated, though. The oldest-known Rocky Mountain juniper grows in New Mexico and has lived 1,889 years. Western juniper (*J. occidentalis*) lives even longer. One individual in the Sierra Nevada was 2,675 years old when it died. In fact, only three species of tree can live longer than western juniper—bristlecone pine, a species of cypress found in Chile, and sequoia.[19]

*

We have seen how ancient people used juniper, and those prehistoric uses continued into historical times: Puebloans continued to use it for their structures; Great Basin groups made wickiups from juniper wood and used bark for thatch; and the Goshutes used the bark to line food storage pits and to cover floors.[20] Paiutes used juniper wood to make bows, digging sticks, houses, sweat houses, and farming tools. They shredded the bark and used it in sandals, skirts, ropes,

The framework for a Navajo forked-pole hogan, Alkali Wash, Yavapai County, Arizona, reportedly dated to 1850. Photo taken in 1953. Used by permission, Utah State Historical Society.

mats, and baskets.[21] The Navajos, who call the Utah juniper *gad bika'igii*, have used the wood to make cradleboards, prayer sticks, hogans, and fences.[22]

In 1849, the U.S. government sent Howard Stansbury to survey the Great Salt Lake. He traveled around the entire lake—no mean feat. Having skirted the northern edge of the lake, his exhausted group, including the hard-working mules, gratefully reached water and pasture at Pilot Peak on October 29. There they noted some dwellings built of juniper.

> In a nook of mountains, some Indian lodges were seen, which had apparently been finished but a short time. They were constructed in the usual conical form, of cedar poles and logs of a considerable size, thatched with

Stockade-type corral fence; Utah Writers' Project photo, location not identified. Used by permission, Utah State Historical Society.

bark and branches, and were quite warm and comfortable. The odour of the cedar was sweet and refreshing. These lodges had been put up, no doubt, by the Shoshonee Indians for their permanent winter-quarters, but had not yet been occupied.[23]

In 1922, while helping to scout out good dam sites on the Colorado River, John Widtsoe came across Navajos also using junipers, but in a different way. Women on the Navajo reservation were weaving rugs on looms "placed as to get shade from the cedars which now cover the country," he wrote in his journal on September 18. "The weaving apparatus is suspended between two cedars." He didn't take note, but the looms were probably also made from juniper limbs.[24]

During his travels through Utah and Arizona, Widtsoe would have passed miles of fences made from cedar posts. By then, fences would have completely

A rip-gut fence in Orderville, Utah, 1954. Used by permission, Utah State Historical Society.

changed the West. In 1934 Cole Porter wrote: "Oh, give me land, lots of land under starry skies above; don't fence me in. Let me ride through the wide-open country that I love; don't fence me in." But "fencing in" had already happened decades earlier. The seasonal migrations of indigenous peoples, the roamings of Spanish Mexican traders, the wanderings of trappers and explorers, and the thousand-mile cattle drives had all faded into history. Enclosure took over: the parceling off of land so that a person could say "mine." You need a fence if you want to keep your livestock from wandering off, to keep animals (wild or domestic) from getting into your fields, and to make sure everyone understands who owns what. Across the West, junipers provided an apt means to that end. Countless cedar trunks and limbs marked boundaries across the West.

In many places settlers used cedars to build stockades. For instance, the settlers at Lowder Spring, north of Panguitch, Utah, enclosed five acres with eight-foot-high walls made from vertical juniper posts set tightly together. To make it more secure, they also dug a deep ditch around the perimeter.[25]

Another kind of fence made of junipers was called "rip-gut." Settlers made these by cutting poles and setting them against each other so that they jutted out in various directions. This made a truly fearsome barrier. It's not hard to imagine how an animal that tried to jump one of those jagged fences could experience firsthand the reason for the name "rip-gut." In Springdale, the little town at the entrance to what is now Zion National Park, each farmer built one of these fences running from the Virgin River to the tall sandstone cliffs.[26]

Into the twentieth century, homesteaders in the remote and severe Cedar Mountain area in Utah's Emery County also relied on rip-gut fences. Though barbed wire had been invented back in the 1860s, ranchers who didn't have cash continued to rely on wood alone. The homesteaders on Cedar Mountain built their rip-gut fences with the abundant juniper and pinyon—free for the cutting. If, as happened sometimes, people from the valley coveted and stole all that cut "firewood," the ranchers would have to cut more trees, haul the logs, and rebuild. It was just one more task that went into the near-impossible endeavor of making a living on Cedar Mountain. By the 1930s the homesteaders had all given up.[27]

In the central Utah town of Minersville, settlers took their rip-gut fence a step further. They were dealing with rabbits, thousands of them invading barns, gardens, and grain stores. Their extermination plan involved first building a rip-gut fence. "Some of that old fence is intact today down on my farm," said a Minersville farmer in 1977. "And then on the outside of this rip gut fence, they cut small cedars and cedar limbs or anything that could be stood on end and made a solid fence right around the whole Minersville. It took women, kids and everything." Because branches are crooked, rabbits could always find a place to squeeze through, so the settlers cut some large inviting openings into the fence and dug deep pits on the other side. As the rabbits skipped through these tempting entryways, they fell into the pits. There they had to wait for the humans who grew those alluring grains and vegetables to come around with clubs. The men killed as many as five hundred each night.[28]

Cedar fences might be strong but they are not indestructible. It depends on how they are made—or what they have to withstand. In the San Juan River area,

> Rye Butt told how one night some cowboys put a bunch of cows they had been rounding up into a small corral to hold them until the next day. Since there were a few wild ones in the bunch, one of the cowboys objected when the boss told the men to tie a couple of night horses to the corral posts. The boss replied curtly, "That corral is made of cedars. I built it myself. They won't get out of there." Shortly after midnight one of the night horses shook his saddle. The rattle of the stirrups and other saddle fittings was too much for the herd, and the cows stampeded. They went right through the cedar fence, taking down several panels and trampling three cows to death as they fled.[29]

But mostly, it seems, cedar fences lasted. Levi Peterson wrote of his childhood in Snowflake, Arizona, that rip-gut fences built in "pioneer times" still stood along the lanes. The fences and the wood that made them were integral to life in the area. "That same aromatic juniper fueled the stoves of the villages," Peterson writes. "Men and boys earned their tickets to the annual wood dance, held on Thanksgiving night, by hauling, sawing, and splitting a winter's supply for the village widows. Sometimes in good weather the entire village repaired to the nearby junipers, ate a potluck supper, and enjoyed songs and orations around a roaring bonfire built of whole trees. I remember one such occasion when a local cattleman, accompanying himself on a guitar, sang 'Home on the Range.' The Arizona sky stretched from horizon to horizon, ablaze with a multitude of stars that modern city dwellers can have no conception of."[30]

*

Brant and Betty Wall spent their early years in intimate association with the western sky, earth, and hard work. When I spoke with them a few years ago, they had lived more than fifty years high on Salt Lake City's east bench, but they had not lost their connection with juniper country. Maybe those years of hard work in the high desert were what had kept them vital into their nineties: Brant still practiced law every day, rode his quarter horses, and cared for land he owned in Joe's Valley.

Brant grew up in Castle Dale, in the middle of Utah on the edge of the Colorado Plateau. A coal-mining and ranching community, Castle Dale sits near the jumbled red rock of the San Rafael Swell and the high spine of the Wasatch Plateau. Hard work "was just part of the deal" on a ranch—irrigating, cutting, mowing, raking, and baling hay; herding and castrating cattle. "Soon as you were old enough to work you were active. You didn't sit around in the house and play with all those fancy doodads. You had chores to do: chop, get coal, feed pigs and chickens, milk cows, throw hay to them. That was the standard routine, every day. You didn't have time to get sore muscles. We didn't think anything of work. It was just a way of life. I can't imagine how people in the city survived the Depression. We had a cow, chickens, eggs. Everybody raised pigs. Everybody had a garden. Except for sugar and spices, everyone was self-sufficient."

A major part of that work was building and repairing cedar post fences to last. Near Brant's town was a bench that the locals called "Cedar Country." There men would cut trees, trim the branches, and cut fence posts to size. After all that work, they would dig the post holes—by hand. This was labor at its most raw—muscles and sinews, arms, legs, hands, abdomen, back, neck, and feet fully engaged in work that you, with your own sweat, had to finish if it was to be finished. I asked Brant how long it took to get a pole set.

> Hell, it depends on the kind of dirt. The longest part of building a fence is digging the hole. You didn't have the fancy post-hole diggers they have today. It was a matter of a crowbar, pick, and shovel. You'd dig an eighteen-inch-wide hole and go down eighteen inches. You'd put the pole in and tamp the dirt down with the shovel handle to set it tight. Some soil will have a lot of gravel and rock; some areas it's swampy clay. In a rocky area, you can get rocks so big you have to abandon the hole and stretch the wire further. You let the posts settle a while before stretching wire. When they're good and set, you bring the wire around the post. You put the claw on the post, and pull and pull the wire around. Then you staple it with a galvanized staple.

If there was a long distance between posts, he said, you might have to bring in a team of horses to stretch the wire tight enough. You'd snug it up against the post and pound in the staples. When you came to a corner post or a bend, you'd put the post in at an angle or brace it to help it resist the tension and keep it from getting pulled over. You'd string three strands of wire usually, unless you were fencing for sheep, and then you'd need four strands to keep them from getting out.[31]

That's how the West was fenced.

*

There was a time when those people who had access to an abundance of junipers could make some money by cutting and selling posts. The people of Juniper, Idaho, did have an abundance—obviously—and they took advantage of it; one resident called the cedars a "godsend" to the people in the valley. To make money, people would cut cedar posts, load them onto wagons, and drive them somewhere to sell them.[32] In the winter Alvin Lund and his boys Harold and Lee wrapped their legs from feet to knees with burlap and canvas strips and went off into the hills to cut posts. Once they had a decent pile, Alvin would load one hundred posts at a time into his wagon and haul them to Tremonton and Ogden in Utah, sell them, and bring back apples or other goods. Another resident traded his posts for cows. When Merrell Nelson was born, his father paid the doctor bill with a load of posts. Lars Anderson said he could sell his posts for ten cents apiece in the nearby small town of Snowville, twenty-five cents in the larger town of Tremonton, and forty-five cents in the city of Ogden—the longest round trip of the three. One year, the brothers Carlton and Guy Wilson camped in the mountains just to cut posts. With prodigious effort over several days, they cut one thousand regular fence posts and one hundred larger posts. When one of the brothers got sick, they had to leave the wood behind for a few days. When they returned, someone had stolen the whole pile—$350 worth of work, they estimated.[33]

Fences made by setting hand-cut cedar trunks and limbs into hand-dug holes still stand in rural regions of the West. ("You'd be amazed," says Brant. "I could show you posts that were there when I was a kid.") The posts, each one unique, with shreds of bark still clinging to them, weather to silver gray as they continue to support the aging barbed wire. Many fences today are sagging from old age and neglect. But mainly they are gradually disappearing. Steel posts and commercial wood posts—not to mention development of one kind or another—are replacing these old fences.

Besides fences, Anglos also used juniper poles for various structures. One, a bridge, figures in local legend. Around 1910 a big timber wolf was roaming all over the Arizona Strip and killing cattle, the story goes. When a group of cowboys decided to get rid of him once and for all, they chased him to the precipitous edge of Paria Canyon. They knew they could trap and kill him there. But they never did. Instead of finding the wolf, they found a sixteen-foot bridge built of juniper poles and rawhide that spanned a narrow part of this deep canyon. The cowboys figured that Robbers Roost outlaws had built it. Like those fugitives, the wolf had apparently evaded "the law" by crossing the bridge north into Utah.[34]

*

Over time, juniper wood has made modest economic contributions, being used for things like fuel, paper, particleboard, and pencils. A lot of the thinking about pinyon-juniper woodlands—and there is quite a bit—has centered on how to increase this economic contribution. Back in 1975, foresters and range managers discussed the PJ ecosystem in a wide-ranging symposium. One presenter noted that the U.S. Forest Service had classified PJ forests as noncommercial. This "immediately puts them into a general 'no-value' bracket. This results in the rather hasty conclusion that pinyon-juniper lands are liabilities." As a result, managers would eradicate the trees in order to put more cows on those lands and produce more red meat. But, the presenter argued, the trees keep coming back after being removed, so why think only of steaks? The woodlands could also be managed for wood products like firewood, fence posts, essential oils, charcoal, particleboard, veneer, pulp, turpentine, and resin. At the same time, cows could still graze on the land.[35]

If there's a potential for economic value in something, someone will find it. People sell lumber, logs, and furniture made from western juniper, a tree found mostly in Oregon and California that can grow tall and relatively straight. And in Texas, the rampantly growing Ashe juniper (*J. ashei*) also provides income. Until the early 1970s, cedar yards sold logs cut by "cedar choppers." This was a rather lowly profession. The homes of cedar choppers, scattered within the cedar brakes, "had no doors, no windows, no screens," one observer said. "The kids' faces were smudged with all the charcoal that they burned in piles, and the children came to school with so much cedar wax in their hair you couldn't comb it out."[36]

Today, companies making essential oil also profit from Ashe juniper wood. Every year, thousands of pounds of juniper oil perfume items ranging from Tide detergent to Irish Spring soap. To make essential oil, companies grind the wood and then use water to steam the oil out. "We grind it up and what's left is the soul of the tree," says Don Baugh, vice president of one of these oil companies. The chip waste goes to make particleboard, mulch, bedding for horses and poultry, filler for crevices in oil well holes, and so on. "I have one fear," Baugh says. "And that's when I die and go to heaven, God would be a cedar tree."[37]

*

Maybe God *is* a cedar tree—and grass, rock, and water. When some people say that God is in all things, they are expressing the idea that the holy is all around us, even in the *materia* we use. I usually don't think much about the holy in "stuff." In day-to-day life, absorbed in facts and utility, I take trees for granted as resources for survival or comfort, or as beautiful decorations in the mountains. But when I make space in my psyche and time to experience the world in stillness, my inner and outer landscape shifts. It's hard to explain. I don't think I have to, though, because surely everyone has experienced that timeless, interconnected feeling. When we were children and everything was fresh to us, we probably felt the aliveness of the world, and so did our distant ancestors. A Halkomelem elder named Lescheem, born about 1871 in British Columbia, was one of the generations of indigenous people who lived in relationship with the alive world. His great-grandson said that Lescheem carefully taught "respect. And that goes for everything.... And he always said, respect the fish, and your elders, your people, including the trees. Even the *sqweyalth* today, the wind. Everything. He included everything. Talk to it. Thank it. They all have their own spirit. . . . So, if you're respectful to everything, they'll do the same."[38]

"If you are respectful to everything, they'll do the same": that's another way of saying that everything is interconnected, junipers included, and that what we do to "everything" will create multiple reciprocal effects. And it's another way of saying what Albert Schweitzer said: all creatures want to live and flourish, and an ethical life must include a reverence for all life.[39] People don't usually think of ethics when they use a pencil, build a fence, or make essential oil. But we could. We could use *materia* with respect, awareness, and reverence. A few years ago, my son-in-law turned a small juniper bowl for me from the trunk of his family's Christmas tree. He shaped it so that the light grain and the dark knots swirl against each other in a pattern unique in the universe. Each of these knots was once a small branch holding tiny leaves. The grain of this wood once carried water and nutrients to living cells. I now hold this bowl cupped in my hand, against my own living cells. I feel the solidness of the wood, the smoothness of a loved one's craftsmanship. I trace the contours of years in the wood grain and in my memory.

William Wordsworth wrote:

> Thanks to the human heart by which we live,
> Thanks to its tenderness, its joys, and fears,
> To me the meanest flower that blows can give
> Thoughts that do often lie too deep for tears.[40]

Sometimes this juniper wood bowl is just another object on my desk. And sometimes I pick it up. I take the time to reverence what it means to my human heart.

7

Smoke

In 1996 in the bright light of a May morning, my parents and I drove up the steep switchbacks of the Moki Dugway out of Mexican Hat, Utah, onto Cedar Mesa and out along an obscure dirt road. We were going to try to find a little-known Ancestral Puebloan ruin tucked under a canyon rim. The hike started out as one of those where you angle down across slickrock and sand, winding through scattered cedars, saltbush, and prickly pear. A lizard or two skittered out of the way. A raven did its acrobatics, ducking into the canyon. We walked and talked. Then the trail turned into a little scrambling. We scrambled down a steep sandy part and got to a shady place, the faintest of seeps staining the pink sand. It was cool here, under a sinuous, rugged, ragged, motherly cedar.

"I'll just wait here," my mother said. I told her she had to come see the ruin. But she insisted. Maybe her knee was bothering her. But maybe not. Maybe she just saw a luminous place and needed to spend an hour there.

My dad and I went on, circling the canyon rim along a sandstone ledge. The ruin stood intact. Thousand-year-old fingerprints marked the mortar between stones. Cedar log vigas roofed the rooms. And the kiva—the sacred ceremonial space dug into the earth—was still covered by a complete roof of cedar logs, brush, and mud.

When we got back to my mother and the juniper, we told her what she had missed. As we walked, she and I fell behind. She told me that something had happened back there, under the tree. It was some kind of experience. She said, "It was about you." I asked what had happened, but she said only that under the tree she had a strong intuition that I would be needing her help soon.

I was skeptical, but I felt a little wave of fear too, because my husband was pretty sick then.

"I don't want to hear that," I said. But she was right. My husband died in September.

At the time, I didn't know that several peoples have regarded juniper as a gateway to the spiritual world. Nepalese Sherpas, for instance, revere certain juniper trees growing in key places on the landscape and build small altars for prayer beneath the trees.[1] For many peoples, though, it is juniper smoke that has symbolized and facilitated that spiritual connection.

*

"The odor of burning juniper is the sweetest fragrance on the face of the earth, in my honest judgment," wrote Edward Abbey. "I doubt if all the smoking censers of Dante's paradise could equal it. One breath of juniper smoke, like the perfume of sagebrush after rain, evokes in magical catalysis, like certain music, the space and light and clarity and piercing strangeness of the American West. Long may it burn."[2]

And long it has burned, hot and steady. I remember my first juniper campfire in the early 1960s, camped east of Six-Shooter Peaks outside Canyonlands National Park. I had never been to the Colorado Plateau, and I think that after experiencing the first day on that wide land, sitting around the juniper fire as night swallowed the mesas and pulled down the stars, might have sealed the deal. I "imprinted" with this landscape like a newborn imprints with its mother, and coming to it would always feel like coming home.

For how many centuries have people smelled juniper incense as they cooked their food and warmed their bodies? The earliest groups moving into the American West thousands of years ago would certainly have made fires of juniper. The fibrous bark would have been effective tinder. Twisted into a rope, tied with yucca, and wrapped into a coil, it could also serve as a "slow match" that could keep smoldering for hours. The Ancestral Puebloan and Fremont cultures continued to use juniper for heat and fuel. They would have burned it in kivas like the one on Cedar Mesa, and in fact, they may have used the smoke as part of their ceremonies, as Navajos and others do today.[3]

Thousands of Euro-American settlers needed juniper fire just as much as those early people did. Homes and families drew warmth for bodies and cooking from the wood that had grown so slowly in the dry and rocky hills. Long after people in the cities had gravitated to the convenience of coal, oil, and electricity, many a rural home kept a glowing wood fire in the cook stove, the cedary scent wafting through the little kitchen.

Forester William Hurst remembered, "In my Father's home in southern Utah, we relied entirely upon [pinyon and juniper] for cooking and heating until 1948. Countless others did also." He went on:

> My Mother would burn only "cedar" in the cook stove, since it was clean and free of tar. On the other hand, pinyon was the standard heating fuel. They had to be carefully separated in the wood box which, for many years, I had the job of filling every evening. Hauling the wood from the foothills in the fall and winter was a carefully planned-for task that required at least a month's time each year. Much of the pinyon was harvested during the cold winter months when it could be shattered with the blow of a pole axe. I vividly recall the shrill screech of the iron-tired wagon wheel as it rolled over the snow on the below-zero morning as we made our way to the pinyon-juniper forest for a load of wood.[4]

Growing up in the tiny town of Lynndyl, Utah, Betty Wall eagerly watched the men bring home big loads of juniper, because in those piles of "beautiful firewood—big old trees that had been there for eons," she loved to create little playhouses. "They'd bring the wood into town and chop it with an ax," she said. "If you wanted to cook a meal or keep warm, you'd start a fire with wood, then put the coal on. The kitchen would have a cookstove—it was all pretty standard. When the times got a little more modern, people put in plumbing. They'd put a water jacket on the stove. The water would go through that and heat up and then go into the tank."

This was the life of a child at that time and place: a hideaway in the woodpile, night games of Run Sheepie Run, potato roasts on bonfires, occasional "shows" at the town church from a projector and rigged-up screen, catching quimps (ground squirrels) and putting them on leashes, swimming in the Sevier River.

"Cedar was great firewood," says Brant Wall, who grew up miles east of Betty. "There's something in it that would burn hot and fast. It burns up completely.

Everybody had a woodpile. You'd go chop wood. In the evening you'd chop kindling and put it by the stove so in the morning you'd have it. We had a kitchen stove and a Heaterola in the living area."[5]

Fred Esplin told me that as he was growing up and herding cows on the Arizona Strip, he used to smoke "barkies." "You'd take juniper bark, rub it between your hands, and roll it in newspaper or whatever paper you had. My Navajo brother and I, when we were gathering the cows, would smoke these and think we were real ranch hands. We didn't inhale, though. It was real pungent blue smoke." Brant Wall also smoked barkies. I asked him how they tasted. "Like cedar bark," he said. When I told this to my dad, he said, "Oh sure. That's what twelve-year-old boys did back then. But cedar tasted a lot better than the other things we smoked." Such as? "Cow pies."[6]

*

Fire has many forms and uses, and native peoples set fires across the landscape as well as within dwellings and campsites. The Catholic priests Domínguez and Escalante and their party, exploring where whites had never gone, found burned areas and saw smoke around Utah Valley. On September 23, 1776, Escalante noted in his journal that

> the pasture of the meadows through which we had been traveling had been recently burnt, and that others nearby were still burning. From this we inferred that these Indians had thought us to be Comanches or some other hostile people, and since they had perhaps seen that we had horses, they had attempted to burn the pastures along our way, so that the lack of grass might force us to leave the plain more quickly.[7]

The Spaniards probably inferred wrong. The Utes likely had set the fires for more important reasons—like survival. Fires could clear out brush and tree sprouts and encourage lush grass growth in the spring—thus improving grazing for big game and renewing the people's seed crops. People could also "corral" and collect insects with a fire surround or use fire to facilitate hunting. The autumn time, when the Spanish passed through, was a logical time to burn; the Utes would have finished their seed collecting for the year, and the cooler temperatures meant that the fires wouldn't get too violent.[8] A hundred years later, John Wesley Powell saw evidence of fire "everywhere throughout the Rocky Mountain Region" and surmised that "in the main these fires are set by Indians."[9]

But did burning by native people affect the pinyon-juniper woodlands? Did this burning keep the trees from expanding into grass and sagebrush land? This is in fact a question that has raised discussion and research over the years. However, it's not easy to figure out the history of fire in pinyon-juniper, and we still don't completely understand that history or the effects of fire. The average person may not care about "fire regimes"—why, how, and how often woodlands burn and what happens as a result—but what we humans do to the PJ ecosystem is based partly on what we think we understand about fire.[10] According to the scientific literature, the current understanding about fire and PJ includes these observations:

- Almost all PJ fires observed since settlement of the West have been high-severity fires that burn and kill all the trees in dense woodlands, though sometimes they leave "islands" of trees alive.
- Although many have theorized that PJ woodlands were in the past kept in check by low-intensity ground fires that killed small trees but left large ones alive, there's not much evidence to confirm this. This might have occurred in some areas, but probably not generally throughout the West.[11]
- Fire scars are crucial to understanding fire history but are problematic in PJ. Sometimes scars are present, but they are not common. And since cross-dating juniper is difficult, it's hard to tell whether two scarred trees burned in the same fire.
- High-intensity fires in PJ occurred only infrequently before settlement. In many places, the pinyon-juniper woodlands have been stable for centuries, affected only by isolated lightning strikes.[12]
- In some places, such as northern Colorado and Mesa Verde, increased fires since settlement—aided by flammable cheatgrass—are causing the pinyon-juniper woodlands to shrink.[13]

The enormous Milford Flat fire of 2007, ignited by lightning, burned hundreds of thousands of acres of land thickly covered with juniper.
Courtesy Bureau of Land Management, Utah.

If a fire starts in hot, dry, windy weather, it can spread quickly through a dense stand of trees and get very hot and explosive—and hard to control.[14] "The shaggy bark of the juniper made fire brands to Satan's liking," one observer of a big fire in 1950 exclaimed. "Flaming strips of this bark, often 2 feet or more in length, were hurled ahead to wrap themselves around other trees which caught fire with a roar and gave off ropelike strips of bark to repeat the process."[15] The volatile compounds in juniper wood give a juniper wildfire extra vigor and staying power.[16] For instance, in July 2007, lightning ignited a wildfire near Milford, Utah. The fire moved fast through the "continuous fuel load" of a thick juniper woodland and became the largest in Utah history to date. Before it was out, 567 square miles had burned. A hot fire like that not only kills vegetation, destroys the seed bank of native plants, and creates an opening for weeds like cheatgrass; it also damages nutrients, organic matter, and organisms in the soil.[17]

After a big fire does occur, the land will adjust in its own way, depending on all the factors influencing it, but in most cases, the pinyons and junipers that were burned will slowly reestablish and become dominant. On Mesa Verde, investigators found that during the two years after a fire, annuals like sunflowers and lamb's-quarters moved in. After four years, perennial grasses like Indian ricegrass, muttongrass, and squirreltail would dominate. After twenty-five years, the site would become brushy with plants like bitterbrush and fendlerbush, and also pinyon and juniper seedlings. After a century, the site might be covered with dense brush along with young pinyon and juniper trees and some grass and perennials. At four hundred years, the trees would be big, and the accompanying plant life would have shrunk—from forty-seven species in the years right after a burn to twenty-six species in a "climax" forest.[18]

*

An explosive pinyon-juniper wildfire gives some sense of the energy embodied in the trees. That energy is concentrated when pinyon and juniper are made into charcoal. The process makes a lightweight fuel that burns twice as hot as wood alone will burn.

In 1850 about 120 men, 30 women, and 18 children set off to settle the Cedar City area of Utah as part of the "Iron Mission." Their mission was to manufacture iron, using the innumerable cedar trees in the region to make charcoal for smelting the ore. However, this charcoal plan changed when the group discovered a seam of coal near the iron deposits. Eventually, the Iron Mission did produce iron, but not of great quality or quantity; the settlers never quite figured out how to solve all the simultaneous problems of settling a frontier community and establishing a profitable industry in an unpredictable, isolated land and society. Many factors contributed to the ultimate failure of the Iron Mission. Morris Shirts names in particular the so-called Utah War, when U.S. Army troops came west to quell any Mormon rebelliousness and replace Brigham Young as territorial governor. The Utah War, he says, "was the last nail driven into the Deseret Iron Company coffin." The Mountain Meadows

Charcoal kiln at Old Iron Town, west of Cedar City, Utah.

Massacre, in which Mormons murdered California-bound emigrants, "could be considered the first armed skirmish of this affair. Over half of the Deseret Iron Company cadre were involved in it, including the leadership. This shattered the spirit of the enterprise." The year after the massacre, Brigham Young called an end to the Iron Mission.[19]

But enterprising men still had an eye on all that iron ore and all the pinyon and cedar in the area. One April day, Eddie and I headed west from Cedar City on U-56, traveling to see where these entrepreneurs had built Iron City, now known as Old Iron Town. The sun was dropping down toward a ridge, lighting a swath of clouds and a strip of cerulean sky. In the southwest, lighted rain streaked down. Cedar trees grew all along the way, covering the folded country, sending roots through pink and rust-colored soils. The hayfields and pastures were turning green. A pinto foal grazed by its mother. The road rose and fell with the landscape. It would have taken a tediously long time to get where we were going with a horse and wagon. At Old Iron Town Road, we turned south for five miles. The low-angled sunlight gleamed across stretches of sagebrush, and shadows were growing in the cedar woodlands by the time we reached the place.

A huge beehive-shaped kiln of lichen-spotted rocks still stands at Iron Town. Inside, I looked up at blackened walls and an opening way up at the top. Outside, a mourning dove called. Somebody—the BLM?—had stabilized or rebuilt the walls using concrete mortar. Little brick-sized openings encircled the kiln at breast height. The workers in Iron Town would have made charcoal by piling juniper and pinyon logs inside the kiln and lighting them on fire. They would have opened and closed the kiln holes in such a way as to keep the wood just smoldering. It took twelve days to produce fifty bushels of charcoal, enough to smelt one ton of iron ore. Another kiln, now in ruins, stood nearby.

These kilns were built around 1868, ten years after the Iron Mission ended. That year, Ebenezer Hanks and a few others revived the industry by forming the Union Iron Company here at Iron City. They wanted their company to be a cooperative, a venture dedicated not mainly to profit but to building a self-sufficient, economically just community. Here an industrious group of people would live together in a little Zion. This was one of those times in history when someone dreamed large and beautiful dreams of a community built on cooperative principles. Iron Town, like many such dreams, didn't ever reach that vision.

Ebenezer Hanks had first come west as part of the Mormon Battalion, one of more than five hundred men who enlisted to fight in the Mexican-American War and use their pay to help build the Mormons' "Kingdom of God." Along with a significant number of other women and children, his wife made the march too, working as a cook and laundress. In the process of marching to California, the battalion established a new road across the country, and then they helped establish San Bernardino. Hanks returned to Utah, became a wealthy merchant, briefly served as the mayor of Provo, built a cotton mill in Parowan, Utah, and later—after losing his shirt at Iron City—helped found the little town of Hanksville, in the bleak landscape of the Dirty Devil River.

In some respects, Hanks and the other founders of Iron City chose their site well: it sat amid pinyons and junipers, close to rich ore deposits at Iron Mountain, and beside a stream that could power a bellows. The Iron City cooperative allowed labor representatives to sit on the board and workers to hold stock. Before long, the town had a blast furnace, machine shop, pattern shop, grinder, engine house, foundry, school, post office, butcher shop, and, of course, homes. The blast furnace chimney remains today, telling a mute tale of industry in a remote outpost surrounded by miles and miles of cedar-covered hills.

Fueled by charcoal made from those cedars, the ironworks ran day and night; at its peak the works could produce five to seven tons a day. Some of the iron went to cast the statues of twelve oxen that still stand within the Saint George Mormon temple. Other iron went to mines and railroads in Nevada. But the stars just didn't align for this company. Unable to sustain itself as a purely Mormon enterprise, the company had to go after outside capital and labor, which brought in outside influences (including the usual mining town extracurricular activities) and stronger capitalist values. Shipping costs, competition from other iron companies, a dearth of disposable cash in Utah Territory, the financial Panic of 1873 with its ensuing depression, and the difficulty of balancing the Mormon economic cooperation ideal with a capitalist-type venture—all held the enterprise back. Finally, in 1876, the whole thing folded.[20]

Ragged old junipers now grow where the bustling town once stood. A sign along an interpretive trail through the trees informs readers that jackrabbits, coyotes, and foxes eat juniper berries. The ruins of a sandstone rock house stand along the trail, slowly slipping into entropy. Pinto Creek, within a wide, deep wash, runs nearby, invisible but for the sound of running water. Over time, other companies have tried to make money from the iron deposits still there; open-pit mines scar nearby cedar-clad hills.

At sunset in April, the sky was gold and mauve, a classic Old West sky. Rocks painted with pale green lichen rested on the earth. Scarlet paintbrush lighted spaces between sagebrush. It was hard to imagine the noise and smoke of Iron City.

*

In 1885, entrepreneurs built four charcoal kilns near the burg of Leamington, in Utah's West Desert. Two of these still stand at the edge of Highway 132. The owners of this enterprise had no mining dreams; instead, they made money from the charcoal itself, hauling it 112 miles to Salt Lake Valley to sell as fuel for the smelters in Murray that fouled the air and sent acid rain on crops. Later, they sold it to railroads, which would use it as insulation to keep ice cream frozen. The charcoal enterprise took dedication, tremendous work, and a whole lot of trees. Merrill Dutson's father made four trips a day into Tank or Wood Canyon to haul a cord of wood each time. According to Dutson, the kilns at Leamington held twenty-five cords of wood each, and it took three or four men around twelve hours to fill a kiln by standing four-foot-long logs upright inside. When they had the kiln filled, they used bricks to close the small vents near the bottom of the structure, and someone pushed a long torch into the kindling at the center of the pile to light the wood. During six to eight days of slow burning, someone had to watch the kiln and open or close the vents to keep the fire at the right temperature. At the end, the pile had to sit for two or three days and then cool for a few more days, with the help of water poured in through the opening at the top of the oven so that the pile would not ignite again. The end product, charcoal, had all the organic matter removed so that it was pure carbon.[21]

In Nevada, during the Comstock Era of intense mining (1859–1890s), people made charcoal in makeshift earthen kilns and belowground kilns as well as in rock kilns. An earthen kiln would start with parallel logs laid on bare ground. These would keep the bottom of the pile ventilated. The charcoal makers would then heap wood on the logs; pinyon made the best charcoal, but they also used juniper, and if other trees were available, like Jeffrey pine, ponderosa, limber pine, and mountain mahogany, they might use those. The men would then thatch the pile with green boughs and cover the boughs with soil before lighting the pile.[22]

Before railroads started delivering coal, smelters used enormous amounts of charcoal. In fact, charcoal was the largest expense in the smelting process.[23] In Nevada, hundreds of Italian and Swiss Italian immigrants known as *carbonari* cut, hauled, and burned the wood using techniques they had brought from Europe. Chinese and Shoshones also did this work—the Shoshones ironically having to cut their traditional source of pinyon nuts in order to survive in the Euro-American world.[24] The whole situation was hard. Even though charcoal makers provided a crucial product, they earned little respect or pay—less than half the wage of mine workers. They lived in crude camps with bad sanitation, worked incredibly hard, and endured racist mistreatment. Mine owners and teamsters exploited and cheated them. And the Chinese and Shoshones were willing to work for less money than the carbonari were, which led to inevitable tensions. In 1879, the carbonari rose up to demand fair prices for their charcoal. The "Charcoal Burners' War" in Eureka, Nevada, left five carbonari dead, shot by government deputies.[25]

Woodcutting for charcoal devastated pinyon-juniper woodlands around the mining centers in Nevada. Only a couple of years after mining began near Eureka, thirteen smelters were using the charcoal from 530 cords of wood per day—or the trees from fifty acres. Per *day.* Beneath the smoke-filled sky lay land stripped of trees. By 1873, ten miles of woodlands around Eureka had been clear-cut. By 1874, you had to travel twenty miles from Eureka to find trees. By 1878, six smelting companies operated sixteen furnaces requiring sixteen thousand bushels of charcoal daily, and the denuded radius around Eureka had stretched to fifty miles.[26]

In 1879 arborist C. S. Sargent noted "the terrible destruction of forest" following each significant discovery of minerals. In 1887 the Nevada U.S. district attorney, Thomas Haydon, reported that hundreds or thousands of woodcutters had been cutting trees belonging to the government around all of Nevada's large mining camps.[27] The cutting pattern radiated outward from the mining centers, intense closer in, and less intense as distance increased. As rail lines came in and trains could haul charcoal easily, the cutting expanded northward.[28] By then, the cutters had already taken out the big old pinyons and junipers, and the Eureka and Palisade Narrow Gauge Railroad was hauling trees less than five inches in diameter. As they provided trees for charcoal, heating and cooking fuel, and locomotive fuel, cutters deforested perhaps 750,000 acres of Nevada's woodlands.[29]

Around the Arizona towns of Bisbee and Tombstone, it was the same. By the 1880s, trees became so scarce around Tombstone that people began digging up roots for fuel. One resident said, "Go out in the treeless valley and look closely on the ground . . . and you will soon discover long, black-looking roots uncovered in spots. Put your pick under these and pry them up, and it is surprising how soon you can load your wagon with the best stove wood of any land." The stripping of trees from slopes led to floods and erosion—which particularly devastated the residents living in the narrow canyon where Bisbee sits.[30]

*

Navajos, who call the Utah juniper *gad bika'igii*, also used the charcoal from juniper, to smelt silver.[31] But the burning of juniper also had more esoteric meaning for indigenous peoples of the American West. Many used juniper smoke as well as the branches to fumigate and disinfect after illnesses, and to protect against diseases. The smoke also helped people feel safe from evil spirits and ghosts. Maybe there is something universal behind this quality, because widely different cultures have turned to the juniper's powers in this regard. To those who put stock in the Victorians' "Language of Flowers," juniper represented succor and protection. Egyptians burned the branches and berries of the juniper as part of purification rituals. During the plagues in Europe, people burned it in the hopes of keeping that lethal disease at bay. French hospitals burned juniper during the smallpox epidemic of 1870, and, in the midst of the devastating influenza epidemic of 1918, which killed twenty to forty million people worldwide, many hospitals hoped that juniper essential oil, sprayed into the air, would hinder the spread of infection.[32]

Dorena Martineau, of the Cedar City band of Paiutes, says that when she was small, "the old people would have an old wood stove and have a cast iron fry pan and put some juniper leaves in the skillet and place it on the stove." Fragrance from the smoke would drift throughout the house, partly for the pleasing smell, and partly "for protection and to ward off bad spirits." Juniper, she says, is important in the Native American church, providing a purifying and sacred smoke for prayers.[33]

Juniper and its smoke have played a part in ceremonies across wide space and time. Some fifteen thousand years ago, someone left behind a small spoon-shaped bowl in Lascaux Cave, the cave complex in France with magnificent Paleolithic-era paintings on the walls. The bowl was beautifully carved from sandstone and had juniper soot in the bottom. Evidence suggests that people used this bowl in rituals as an incense burner, to produce aromatic smoke. Alternatively, it could have provided light in the cave.[34] The Sumerians and Babylonians burned juniper for their gods; it was sacred to the goddesses Inanna and Ishtar of Mesopotamia.[35] Celtic people believed the smoke could aid clairvoyance, and they used it to contact the Otherworld at the Samhain festival in the autumn—the beginning of the Celtic year. In central Europe, juniper smoke helped people clean and fumigate in springtime, and they believed it protected against witchcraft. Farmers carried smoldering branches around their fields to protect their livestock. Mongolian shamans use juniper smoke to raise the "wind-horse" that can transport someone to the spirit world;[36] and a *National Geographic* article describes the burning of juniper twigs by a shaman in Mongolia to attract guardian spirits, who are drawn by the fragrance of the smoke.[37]

Ethnobotanists have recorded that on the American continent the Hopis use juniper charcoal to make ceremonial body paint. Tewas use the bark to ignite a ceremonial fire for the new year. Besides making prayer sticks from the wood, Navajos burn the leaves and use ashes to blacken the body in certain ceremonies, and they use juniper to protect from enemies and witches.[38] They use juniper wood struck by lightning to make the fire drill used in the Night Chant ceremony; they use the bark as tinder. On the last night of the Mountain Chant, the bark is used in the Fire Dance and then ignited. Navajos also make juniper into charcoal by smothering smoldering juniper coals with earth and leaving it overnight. They then grind it into powder that will be the black coloring in sand paintings.[39] "Do not walk alone at night or evil spirits will bother you," says the self-proclaimed "Unofficial" Central Navajo Nation website. Among its list of traditional Navajo taboos is this note: "If a Navajo must travel in the dark, juniper ashes smeared on the face might help, or juniper berries in the mouth."[40]

Smoke symbolizes juniper's intangible qualities, which for some people played a part in life from birth to death. Hopis rubbed ashes on their newborns and made the baby's bed and coverlet of juniper bark "to make him strong and healthy." They held misbehaving children in a blanket over a smoldering juniper fire to calm them down.[41] At death, some groups used juniper to purify and protect. The smoke would fumigate the house after someone had been ill or died. Members of the household might drink juniper tea for several days after death. Water boiled with the branches would be used to wash down the house and furnishings, to wash the deceased person's bedding and clothing, and to wash members of the household. Especially, men returning from burying the body would wash themselves. The wash water might be splashed along the outside of the house and trails leading from it to prevent the dead person from returning. Branches might be laid with the body to keep the ghost from harming or scaring the living. When a Navajo passed from this earth, according to Peattie, "the two men who attended [the deceased] to the last back carefully away from the grave, sweeping their tracks with Juniper boughs so that deathliness shall not follow them from the grave."[42]

Ethnobotanists at the Arizona Ethnobotanical Research Association actually encourage people to connect with the spirit of juniper smoke. "As herbalists, our most important use of juniper is as a daily purification smudge," they write.

> Many customers come into [our shop in Flagstaff] and comment on the smell. Most people love it. Some people remark that they haven't smelled anything like that since their party days in college!!?!! Often, Navajo children who come in like the smell because it reminds them of their grandma's Hogan.
>
> You can collect your own juniper for smudging. After making an offering, harvest small tips of branches of juniper and let them dry. Every day, say your prayers and burn a small amount in an abalone shell, letting the wonderful smell of smoke fill the air, opening the window to release any negative energy. This is an ancient tradition, which helps to bring all of our prayers up to the heavens. At night, you can do the same thing to say goodnight to the moon, and say thanks for another blessed day on earth. Then you too can know why magical juniper is the Tree of Life.[43]

*

There used to be—that is to say, four hundred years ago—the term "juniper lecture." If you received a juniper lecture you received a severe, pungent reprimand. An author in 1742 noted, "When Women chide their Husbands for a long While together, it is commonly said, they give them a Juniper Lecture; which, I am informed, is a Comparison taken from the long Lasting of the Live-coals of that Wood."[44]

The coals last long, whether within a hogan, where a medicine man tends a small fire and the smoke blesses and connects the people to the spiritual world; within the furnace of a smelter; in a cookstove; in a campfire on a remote mesa.

The juniper has burned in the cedar cigarettes of kids, in the hands of a man purifying his home, in intentional landscape-management fires, in huge wildfires that crews struggle to contain.

Before it burns, a tree may have lived many years, or just a few. In flames, the tree releases the carbon it has formed and stored throughout those years. The solidity of it becomes smoke and ashes. The scent of its life—Abbey's "sweetest fragrance on the face of the earth"—wafts or rages through the air. The burning may help or hurt. A fire is the ephemeral emblem of tree–human connections and the exchange of energy: spiritual or practical, restful or striving, profitable or devastating. In any case, juniper wood burns hot and clean.

8

Spread

One April some twenty-five years ago a little group ranging from age eight to sixty-eight backpacked Owl and Fish Canyons in southeastern Utah. It's not the most well-known spot in the West—thankfully. We were pretty much alone in the canyons. At the end lay a sandy, relaxing camp spot—almost a resort after miles of hot trudging and boulder scrambling. The next morning, we climbed out of Fish Canyon back up onto Cedar Mesa. The climb was steep, rocky, and tricky enough that we high-fived with joy when we reached the top and took a picture on the rim. But we still had a pretty long walk to get to the car at the trailhead. We headed off through a juniper woodland, following a faint trail. As we wound deeper and deeper into the trees, cedars closed in on every side, ahead and behind and all around. A weird feeling of total disorientation crept over me. There in the trees you couldn't see the horizon, no butte or mesa to steer by. Just trees. If you got off track, you might wander around forever, or so it seemed.

Today around the West thick cedar woodlands are causing what you might call disorientation of a different kind. In many places, young junipers have sprung up into dense stands. This situation has caused alarm among land managers, ranchers, and others; it has given numerous researchers puzzles to sort out; it has sparked heated discussions; and it has inspired various "fixes."

Where are the junipers spreading or infilling, and where are they not? Why is this happening? What are the effects? What should land managers do about it? Whose voices should they listen to the most? If they thin the junipers, what will be the effects? There are no simple answers to these questions, because

A backpacking group on the thickly wooded Cedar Mesa.

these questions get to the heart of our endlessly interwoven relationships with the natural world. We act; the environment reacts in ways not entirely predictable. We react. The environment reacts in ways not entirely predictable, because at the same time, other forces are pulling other strands of the ecological web. Nothing stays the same. "Dynamic disequilibrium" is the norm. What we see happening today is only the latest frame in an "ecological film strip hundreds of miles long."[1]

To figure out what is happening now in any one spot, it's enlightening to go back several frames and see what, if anything, has changed. On Cedar Mesa, which stretches between Elk Ridge and the San Juan River in Utah, junipers are thick now and were also thick 150 years ago (there's a reason for the name of this mesa). When advance scouts for the Mormon "Hole-in-the-Rock" pioneers came through here in 1880, they got disoriented and lost in the cedars until they found a high spot. The four men were truly on the brink of starvation when they reached this pointed little hill. Here, they could look around over the choppy, rocky, cedary landscape and see in the distance the Abajo, or Blue, Mountains. George Hobbs recognized them—he'd been in the area before—and now knew where they were. Hobbs called the hill they stood on Salvation Knoll.

George Hobbs was just twenty-two at the time. But he had already had plenty of experience in pioneering. With his family, he had left his home in England, taken a ship across the Atlantic, walked across the continent to Salt Lake City, and then gone to the edge of the world to settle the little outpost of Parowan, Utah. Now, in order to help guide the settlers lumbering in wooden wagons toward the Four Corners area, he scouted across country that cannot be adequately described—only experienced. The group's job was to find a way from the Colorado River across "the roughest country you or anybody else ever seen; it's nothing in the world but rocks and holes, hills and hollows," as pioneer Elizabeth Decker reported. And she noted this roughest country *after* the group had gotten their wagons down through a harrowing slot in the towering cliffs above the Colorado River (now known as the Hole in the Rock). Today, it sure seems that the endeavor of crossing that country was madness—but by the time the settlers had descended the Hole in the Rock above the Colorado River, they had passed the point of no return, and continue they must.

At one spot, the men scouting out the rough country ahead decided there was no way wagons could descend the sheer face of Gray Mesa. But Hobbs followed what he called a llama down the face, trying to rope it because he thought it was too pretty to kill. The llama (a mountain sheep, no doubt) led him from shelf to shelf until it reached the base of the cliff. Hobbs came back and told his friends there was a way. When the main group arrived, they spent a week cutting dugways from ledge to ledge so their wagons had a "road" to negotiate, but they did it.

Later, when the settlers got into the same cedars that had bedeviled the scouts, the men had to make a twenty-mile path for the wagons by chopping down uncounted tough old junipers in the way. And this was one of the easier spots on this trek. Amazingly enough, nobody died during this journey. Everybody got to what is now the tiny desert town of Bluff, Utah, and seeing a place where a group might eke out a living, the main party decided to stay there, not wanting to travel another inch. George Hobbs himself didn't stay in Bluff but settled in the somewhat more verdant town of Nephi, Utah, had eleven children (with just one wife), became a builder, built a plaster mill in Sigurd that still operates today, and, perhaps because he worked as a carpenter with nails held between his lips, developed and died of cancer of the mouth.[2]

We learn from the reports of explorer Edward Beale that pinyons and junipers were growing so thickly in one area of Arizona that they diverted his 1857 expedition. Beale writes of encountering a land cut by ravines and "covered with a thick growth of pine and cedar trees, and apparently this country extended for a considerable distance." He decided not to hassle with getting through the trees, but to go around.[3] Beale, who packed a lot of adventure into his years on earth, was on this jaunt across Arizona at the request of President James Buchanan, who asked him to survey and build a road from Fort Defiance, New Mexico, to the Colorado River. At various other times he worked as a seaman, naval officer, spy, Mexican-American War hero, exploring companion to Kit Carson, millionaire California rancher, Indian affairs superintendent, and U.S. ambassador to Austria-Hungary. The wagon road he etched across the Southwest in 1857 would become the general course of Route 66, now I-40.

Other expeditions in central Arizona also struggled through thick PJ stands. Joseph Christmas Ives, who had explored up the Colorado River from its mouth in a paddle-wheel steamboat in 1857 and 1858, left the river at the mouth of the Grand Canyon and headed toward Fort Defiance. As he traveled through the upper Verde River watershed, he noted that "a thick growth of cedars and pines offered occasional obstructions to the pack animals, who would get their loads tangled among the low branches." Farther along, he wrote:

> The face of the country continued much the same. The trees generally intercepted the view. . . . At the end of ten miles of weary travel, a steep ascent brought us to the summit of a table that overlooked the country towards the south for a hundred miles. The picture was grand, but the cedars and pines kept it shut out during most of the time.[4]

Where both Beale and Ives described thick woodlands, the trees still grow densely, as they did then.

Near Ash Fork, west of today's Flagstaff, Arizona, A. W. Whipple of the U.S. Army Corps of Topographical Engineers found abundant, large juniper trees in 1854. He wrote, "For fuel it is excellent and even for railway ties it is doubtful whether this country west of the Delnorte can furnish anything superior." Since Whipple had been charged to find a route for a railroad, it's no wonder he saw railroad ties in the trees. What you need from the land is what you see. Somebody must have agreed the trees could be useful, because today a lot of old stumps pepper this landscape, and youthful junipers have sprung up to take the place of those big trees.[5]

Although some places have been thick with junipers since Euro-Americans arrived, in other places the woodlands have expanded and thickened in the past several decades. Whipple also described an area in the present Kaibab National Forest as a "level country, containing prairies mingled with copses of piñons and cedars." Today, junipers crowd the area he called prairie, filling in all the available open space.[6]

In Oregon, surveyor John W. Meldrum wrote in 1870 about a rolling landscape covered with grasses and widely scattered juniper trees. These would probably have been western juniper, or *J. occidentalis*, which flourishes in eastern Washington and Oregon and parts of Idaho, California, and Nevada. Now, however, young trees have crowded in and covered the areas he wrote about, at the rate of fifty to one hundred per acre. Where once the land lay "open, sparse, and savannah-like," with old junipers growing mainly in shallow soil and on rocky ridges, the trees have now pushed into deeper soils, grasslands, and sagebrush steppe. Likewise, when in 1902 David Griffiths of the U.S. Department of Agriculture toured rangelands in Oregon and Washington, he noted scattered stands of juniper on Steens Mountain in southeastern Oregon. Now, in several places young juniper trees have replaced the aspens that grew abundantly there.[7]

In areas with springs or groundwater, spreading western junipers can intercept the water and decrease spring flows.[8] The U.S. Forest Service estimates that since 1934 the acreage of western junipers in eastern Oregon has increased from 1.5 million acres to over 6 million acres. Since a western juniper can use up to thirty gallons of water each day, "increased juniper dominance has been implicated in the desertification of Oregon's rangelands."[9]

A Sierra juniper near Donner Pass, Sierra Nevada.

When Meldrum and Griffiths toured Oregon and noted the widely scattered trees, they were looking at old-growth trees. Though restoration projects (juniper removal) have attempted to solve legitimate concerns, they have, along with illegal cutting, also destroyed many of these magnificent old western junipers, which can grow much larger than the junipers of the Great Basin and Colorado Plateau.[10] A subspecies of western juniper—Sierra juniper—grows only in the mountains of California and western Nevada. One Sierra juniper, known as the Bennett Juniper, towers to eighty-seven feet and has a robust ground-level diameter of twenty-two feet. This massive tree lives in an alpine meadow high in the Sierra Nevada, where a "perfect storm" of ideal conditions has helped it thrive for perhaps three thousand years or more.[11]

People who love big trees love the Bennett Juniper. But the Bennett is a celebrity among junipers, and, just as in human society, millions of "ordinary" junipers don't get special notice or protection. In fact, they get taken out.

*

When Fred Esplin took his father to the Arizona Strip in the early 2000s, it had been sixty years since the elder Esplin had herded sheep there. As they drove around the land and flew over it in a helicopter, the old rancher was stunned to see how much the junipers had taken over the range.[12] This was yet another place where the pinyon-juniper woodland (a name that is often used generically to include both trees, only junipers, or—less commonly—only pinyons) has expanded or thickened. As we have seen, in some places the woodland hasn't changed much, and in other places it is actually retreating, dying off because of drought, insects, or disease. As we have discussed, drought affects pinyons first, and a major drought that occurred in the Southwest during the 1500s might explain why most pinyons there are less than four hundred years old.[13]

The differences among woodlands shouldn't be surprising when you consider that the pinyon-juniper woodlands are as diverse as they are abundant. Every situation—terrain, soils, climate, flora, fauna, disturbances—is different. On the broadest classification scale, we can distinguish between three major types of pinyon-juniper woodlands in the West:

1. *Persistent woodlands* grow in those places where conditions favor pinyons or junipers or both. You might find these woodlands—dense stands with older and younger trees as well as dead snags—in rocky places with poor soil. They're common on the Colorado Plateau, in much of the Great Basin, and in central Oregon, southern California, and central Arizona.
2. *Pinyon-juniper savannas* form where soil and climate suit both trees and grasses. For instance, in parts of northern New Mexico, where most of the moisture falls as summer rain, widely spaced trees may grow on gentle, grass-covered land.
3. *Wooded shrublands* form where local soils and climate suit mainly shrubs like sagebrush, but where trees can also flourish during moist conditions or periods of no disturbance. In these areas, also found in the Great Basin and on the Colorado Plateau, most of the moisture comes in the winter and falls on deep soils.[14]

These are only broad categories. In Arizona and New Mexico alone, more than seventy subhabitat types have been identified. The soils have a lot to do with these differences—igneous, sedimentary, or alluvial; shallow to moderately deep; coarse or rocky; sand or clay; fertile or not.[15]

These differences may help explain why, in each of the main habitat types, pinyon and juniper have increased dramatically in many places and in other places they have declined or stayed the same. Repeat photography helps tell the tale. Many nineteenth-century photos of landscapes in the West show widely spaced trees or open stands. In recent photos of these same places, the trees often lie like dark cloaks over mountains and hillsides. In a few photographs, however, the trees have decreased or have not changed much.[16]

Why does this matter? Who cares about whether junipers were denser a thousand years, a hundred years, twenty years ago? Actually, land managers care a lot. They have to, because of their stewardship over the land and the needs of various stakeholders. Some people get very passionate, and politics, profits, science, trial and error, and clashing values all converge around the issue of juniper densities. As a result, junipers have at times been uprooted, burned, poisoned, and pulverized; they have also been ardently defended and protected. Ranchers

Historical and repeat photography showing the spread of juniper and diminishment of sagebrush communities at Harris Point, west of Kanab, Utah. Utah State University photos; courtesy Bureau of Land Management, Utah.

despise juniper when they think it's taking over rangelands. Fire specialists fear catastrophic hot, unstoppable fires from so many trees filled with volatile oils. Environmentalists litigate against plans to take out woodlands that include old-growth pinyon and juniper. Range scientists and foresters discuss whether junipers are invading other ecosystems, and hydrologists discuss whether junipers affect water flows. Stakeholders gather in conferences and give papers about the trees, their qualities, their spread, their effects on ecosystems, and how human activities affect them. Several decades ago, a forester remarked that he could remember no other species that had been discussed so much in symposiums (with so little resulting action, he felt at the time) as pinyons and junipers.[17] Since then, the symposiums and conferences have multiplied. Science keeps revealing more and more. What is happening, and why? Finding the answer to juniper questions is an unfolding, ongoing process.

*

One early person to address these questions was Aldo Leopold. That would be *the* Aldo Leopold, prime mover of the conservation movement. Leopold wrote passionately and memorably about the natural world and has influenced generations since. "Like winds and sunsets, wild things were taken for granted until progress began to do away with them," he wrote in *A Sand County Almanac.* "For us of the minority, the opportunity to see geese is more important than television, and the chance to find a pasque-flower is a right as inalienable as free speech."[18]

Smart, curious, and resourceful, the young Leopold constantly explored the world around him. He went on to devour books and scientific knowledge about nature and graduated with a Master of Forestry degree from Yale in 1909. His readers know that he bought, restored, and wrote about land along the Wisconsin River during the 1930s. But long before he was rambling about those woods, springs, and prairies, he joined the infant U.S. Forest Service and set off for Arizona and New Mexico.

Out in the West Leopold got the strange idea that it would be a good thing to set aside land—roadless land—as wilderness. At his urging, the Forest Service did that very thing, years before the Wilderness Act came into being. He also looked around at the bare, gullied soils of much of Arizona. Erosion, he saw,

Aldo Leopold on his horse Polly, Carson National Forest, New Mexico, in 1911.
Courtesy Aldo Leopold Foundation, www.aldoleopold.org.

"eats into our hills like a contagion, and floods bring down the loosened soils upon our valleys like a scourge. Water, soil, animals, and plants—the very fabric of prosperity—react to destroy each other and us." Ranchers told him that good grass used to grow on those hills and valleys, and an acre once could support thirty cows. Now the land was choked with brush and could hardly feed one. Leopold searched for answers. To him, it looked like the land had suffered an extended drought, but he couldn't see any evidence of that in tree rings.

In his travels around Arizona, he counted tree rings, mapped erosion, looked at fire scars, and examined plant succession patterns. In one area near Prescott, Arizona, he found here and there the stumps of big old junipers. Young brush and sparse grasses now grew on the site. This situation provided evidence for Leopold's emerging "grass-fire-brush-erosion" theory. He decided that thick grass had once grown there. Fires started by native people or lightning had spread lightly through the grass, killing shrub and juniper sprouts but not necessarily killing big trees. Afterward, the grass quickly grew back. However, when ranchers and their livestock moved in (pushing the native people out), their cattle and sheep grazed the grass to nubs and trampled the roots, destroying the sod that held the soil in place. Grass fires could no longer spread. Brush choked the land. Topsoil eroded, and gullies formed.[19]

In a nutshell, Leopold wrote in the early 1920s, fire suppression and overgrazing combine to encourage brush and trees, and to erode the soil. What Leopold saw and surmised in Arizona had a large influence on theories and juniper management around the West, as we will see. Since his time, scientific knowledge has increased, and in recent years the science has become more nuanced. Although his theory seems right in some places, one size does not fit all; new evidence sheds light on past assumptions. But though details may vary, Leopold spoke true when he warned that humans compromise the land's immune system through careless grazing, plowing, monoculture, logging, mining, and development. The soils and ecological relationships built over centuries or millennia can be swept away in a few years.[20]

As H. L. Bently, an agricultural agent in Texas, described this carelessness in 1898: "In a short time every acre of grass was stocked [with cattle] beyond its fullest capacity.... The grasses were entirely consumed, the very roots were trampled into the dust and destroyed."[21] New Mexico agricultural agent E. O. Wooten wrote in 1908, "The stockman could not protect his range from himself, because any improvement of his range was only an inducement for someone else to bring stock in upon it; so he put the extra stock on himself."[22] And more recently, C. W. (Cedar Whacker) Wimberley of Texas exclaimed, "Mother Nature is a modest girl! You can't denude her soil or she'll cover it with something that cattle can't eat and man can't use. Mother Nature is covering her nakedness where man can't control it." Wimberley was talking about the Ashe junipers (*J. ashei*) growing up where the land had been overgrazed. He had no use for the young junipers and their ways—multiplying, hogging water, keeping grass from growing, sucking soil dry, polluting streams with runoff—even though he himself had once made a living from cutting old junipers for wood and running a cedar yard from 1937 to 1957.[23]

Aldo Leopold would say that in cases like this westerners were "eating up the interest and principal" of their lands. Individual choices like putting too many cows on the land—based on short-term profit and a view of nature as a commodity—add up, and the land suffers.[24] But we're all in on this. It's not just ranchers who overgraze. Everything we use in any way comes from "nature." In the insulated lives of most of us—roaming as we do through neighborhoods, offices, shopping centers, city parks, national forests, freeways—we don't see most of the effects of our actions. One writer notes that when you or I buy a gold watch, we might realize the true environmental (if not human) cost only if we ourselves had to store the mining waste. Imagine a dump truck full of seventy-nine tons of rock laced with acid and arsenic driving up to the house and dumping it on the lawn, he writes.[25] Multiply the effects of that one little watch by all the things one American uses and consumes, and then consider our cumulative effect on the earth. The relationship between grazing and spreading junipers is only one example of what happens when we change one part of the whole.[26]

*

Leopold suggested that fire suppression has aided the expansion of junipers. The theory applies in some situations, but many managers have relied on it as if it applied everywhere. Recent research suggests that persistent woodlands simply didn't burn very often historically, and not at all with the low-intensity fire Leopold talked about. In shrublands and grasslands/savannas, scientists are still trying to determine the historical behavior and effects of fire. But of course there is no one "brand" of shrublands or grasslands; the grasslands of eastern New Mexico differ from those in western New Mexico, which differ from those in Arizona. Ecosystems and the forces that affect them are complex and unique.[27]

Another hypothesis about the spread of junipers has been proposed: they are spreading as part of a natural range expansion that began long before Euro-American settlers arrived. Pollen deposits and woodrat middens suggest that junipers and other low-elevation conifers have been expanding in range since the glaciers retreated. As we have seen, during the great Ice Age, the Utah juniper was confined to low elevations in the Southwest and Mexico, but in the last several thousand years it has moved northward as far as Wyoming and Montana. In western Colorado, it has been shown that more localized juniper infill and expansion began more than fifty years before the arrival of settlers and their cows.[28]

And then, there's another, more recent, influence on juniper range: "Climate change is changing everything," says biologist Jayne Belnap. "Brush [a term that to her includes junipers] is expanding everywhere, not just where there is grazing." Belnap was explaining biological soil crusts to a group of visitors on a hot September day in the backcountry of Canyonlands National Park. The group stood in a unique section of the park, a place rimmed with red-rock cliffs. No cow could ever wander or be hauled into this area—except, perhaps, with a system of cranes and pulleys. Belnap is an expert in biological soil crusts—BSCs.[29] These communities of organisms are a humble but key part of arid and semiarid ecosystems around the world. We would probably never think twice about these dark crusts on the soil if people like Belnap hadn't educated us, mainly by educating federal agencies. Today, you see signs at national parks and other sites encouraging you not to step on the crusts. One particularly memorable sign casts park visitors as a kind of King Kong: a bunch of bumpy little creatures look up in horror at a gigantic boot about to crush them.

The biological crusts on the Colorado Plateau are knobby colonies of cyanobacteria (blue-green algae), lichens, and mosses. They fix nitrogen, prevent water runoff and wind erosion, help build soil, harbor seeds, and in general help the desert blossom—if not as a rose, then with diverse desert plant and animal life. For Belnap, these crusts are the foundation of life on the Colorado Plateau. "Whenever we pull on the thread of what makes the system tick," she says, "we end up with soil crusts on the other end."[30] But the crust is fragile, and boots, cows, or tires can destroy it in an instant. When this happens, the cyanobacteria will come back fairly quickly. The lichens may take several decades to return, and the mosses perhaps 250 years to get fully reestablished. So the robustness of the crusts in this part of Canyonlands National Park—spotted with yellow, orange, and lime-colored lichens and, when wet, abloom with green mosses—is quite exciting if you are a little nerdy about these kinds of things. Here, free from trampling, the crusts have helped create a diverse area of grasses, junipers, pinyons, shrubs, and forbs.

However, Belnap believes that climate change is a force that this area cannot withstand. She says increased carbon dioxide in the atmosphere is causing the expansion of woody plants, which can take better advantage of the carbon dioxide than grasses can. Grazing may exacerbate the trend, but in her mind it isn't the main cause of shrub expansion. "In twenty years," she says, sweeping her hand to take in the desert meadow, "there will be no grassland here. Just shrubs. I think shrubs are going to take over the universe as we know it."[31]

*

In some cases of "expansion," the pinyon-juniper forest is merely returning to where it once grew. Take an imaginary visit to Shoofly Village, in the high country of central Arizona, and look around. If we could visit the village as it was a thousand years ago, we could walk among dozens of rock structures, including an impressive twenty-six-room building. We could visit with people tending plots of corn, beans, and squash; cooking; making pottery, arrow points, and twine; shouting at their kids, rocking babies, or chatting; playing games. We might see turkeys running through the village or hunters bringing a rabbit or duck home. As we looked around we could see a lot of grasses and perennials like lamb's-quarters and ragweed around the village, along with scattered junipers. Most of the time when people lived in Shoofly Village, they kept the juniper population down as they cut it for fuel. But near the end of this village's occupation, as the cutting lessened and as a century-long drought gave junipers an advantage over grass, the junipers began to grow more densely. Today, more juniper trees grow around the ruins than grew around the living village. The area is reforesting naturally.[32]

Nevada provides a large-scale case study in reforestation. According to John Muir, who visited at least eleven Nevada mountain ranges, pinyon grew on nearly every mountain in the state from the base to eight to nine thousand feet. Other people—including scientists, an attorney, explorers, and mining officials—described the PJ forest as widespread, continuous, and dense in the nineteenth century. As we have seen, the mining industry used vast numbers of trees, stripping much of this woodland from areas close to the mines. Today, the young pinyons and junipers that seem to be invading or encroaching on new territory may simply be repopulating places where they were cut long ago. Although for decades many believed that pinyons and junipers have been invading downslope into areas where they don't belong, in at least some parts of Nevada, past descriptions and present analysis suggest that they are simply reforesting. New young juniper stands could be recovering from any number

Repeat photography of Pine Valley Mountains in Washington County, Utah, showing trees repopulating areas where they were chained in the 1960s. The BLM re-treated these areas in 2005; at present, the junipers have again returned. Utah State University photos; courtesy Bureau of Land Management, Utah.

of disturbances: past fires, cutting by prehistoric or historic people, or chainings and other removal projects.

*

"Encroaching." "Invading." These terms are often used to describe young junipers growing into grasslands and sagebrush lands. They carry overtones of theft and attack. These words make junipers the enemy of rightful occupants of the land. To environmentalists, these words are loaded, used by grazing interests when discussing projects to bring back rangelands. Some people use the more neutral terms "infill" and "expansion." But either way, if junipers grow somewhere, they are just doing what plants do, growing where conditions are right—and in many places conditions seem to be getting more and more right for junipers.

The interesting thing is, several of these conditions that could be contributing to increased junipers arose around roughly the same time: overgrazing, fire exclusion, more carbon dioxide in the atmosphere, warmer and sometimes moister conditions, and an end to extensive cutting.[33] Each area where junipers have increased—or decreased—likely has a unique combination of factors contributing to the situation. Scientists and land managers are increasingly realizing, then, that one size does not fit all in juniper management, even though in the past juniper woodlands were often managed as though they were all the same.

Given that in many places dense new stands of junipers have sprung up, land managers have felt and still feel environmental and political pressure to do what they can to "fix" the situation. Over the decades, the question of juniper spread and what to do about it has generated controversies, animosities, and divisions. Studies, hearings, special interest groups, invectives, lawsuits, Environmental Impact Statements—all of these circle around where junipers "should" or "should not" be.

9

Relationships

"The history of the Piñon-Juniper is a documentation of humanity's changing relationship with its environment," said Chip Cartwright, onetime regional forester in the Southwest. "It starts with a relationship in which human fate was determined by the bounty of its un-manipulated environment, and extends through relationships in which people dramatically modify their environment to meet the needs and wants of growing populations.... We understand no ecosystem is a simple thing, but [the pinyon-juniper ecosystem], with so many links to people and cultures, is particularly complicated."[1]

*

Decades ago, two women lived in the little Idaho town of Juniper, members of a particular ecological community. Each of them had a different relationship with that community. A woman identified as Jessie said:

> Picture, if you can, the splendid solitude of that beautiful valley, with its wide expanse of sage and cedar... and to the north, south, and east, exquisite beauty as far as the eye could see, the cooling mountain breezes, the smell of rabbit brush after a rain, the song of the meadowlark!
>
> The unsettled valley was indeed beautiful; wild flowers everywhere! First to come in the spring was a little white growth called Indian biscuits, then came the Johnny-jump-ups, or tiny yellow violets. They were followed by gorgeous bluebells and buttercups; then came Indian paintbrushes, Canterbury bells, wild flax, yellow snowballs, sand lilies, lady

slippers, daisies, sago [*sic*] lilies, yellow dock, prickly pear cactus, slippery elm, sunflowers, sweet Williams, and many more.

Birds were here in abundance! Blue birds, meadowlarks, mourning doves, swallows with their little mud nests, and sage chickens with their wild little broods. There were many little grey birds somewhat smaller than a sparrow, called chippy birds, and also ground owls and nighthawks.

On the other hand, a woman named Winona said:

Mother often said how desolate and barren the land looked with cedar trees, sagebrush, and wide open space with homes quite far apart. The sound of coyotes, hoot owls, and other animals filled the night air.... The mountains were high and rugged with cedar trees, wild bushes and wild flowers growing abundantly.[2]

There's a *Far Side* cartoon of the "four basic personality types," with different people looking at a glass of water.

The first one exclaims, "The glass is half full!"

The second says gloomily, "The glass is half empty."

The third puzzles, "Half full... No! Wait! Half empty!... No, half... What was the question?"

The fourth scowls and says, "Hey, I ordered a cheeseburger!"

I understand Winona's mother. Jessie was probably remembering a carefree childhood relationship with nature, while maybe Winona's mother, burdened with trying to keep her family alive and well, could feel only desolation in her surroundings. Or maybe she couldn't let go of the kind of life she'd "ordered" but not received. "What you see and what you hear depends a great deal on where you are standing. It also depends on what sort of person you are," wrote C. S. Lewis—about a character who could not perceive the miracles in nature.[3] Because of our differences, some people can look at a dolphin, say, and see a smiling animal that they hope they can swim with. Others might see a minesweeper, interesting scientific questions, entertainment, profit, grace, or meat. We routinely differ—and debate—about the relationships between human needs and the natural world.

When it comes to junipers, the fighting is over their removal from public lands: "treatment," in the language of the removers, or "deforestation," in the language of their opponents.

People argue about this issue in public meetings, private meetings, the courts, and the media. These conflicts are one incarnation of long-standing uneasy relationships among different "sorts of persons": county commissioners, conservationists, federal employees, ranchers, wilderness advocates, energy company executives, scientists, rural and urban folks, westerners and easterners. The federal government is often in the middle of it all, since people on both sides of arguments tend to distrust the government and any plan that comes out of a government agency. An aggressive "trust destruction industry" of media and politics fans the flames of hostility, according to Mark Brunson, a professor of environment and society at Utah State University. Media has become politicized, says Brunson, and as any aware citizen knows, politics has become largely about sound bites and media exposure.[4]

*

Long before political relationships—which is to say, before any humans came onto the landscape—webs of interdependent relationships formed unique and complex ecosystems. The pinyon-juniper forests are not just trees, and they are not all the same. Nor does a forest stand alone. As an interpretive sign on Casco Bay, in Maine, expresses it:

> Plants and animals function together as an ecosystem, sharing the earth's raw materials and the sun's energy to do their work before passing these things on to others. Actively involving even the bedrock and the atmosphere, these ecosystems blend into and influence each other so much that our earth is really one great interdependent ecosystem. Just as the shore community below merges into a deep water system and then joins the islands of the bay, we are connected to those islands, the oceans, and the continents beyond.

You could express the human involvement in this interdependency with the fairly recent term "socioecological systems." This term points to the understanding that human systems affect every part of nature, and nature affects every

Pinyon pine (left) growing near a juniper, Capitol Reef National Park.

human system. It's another way of saying we are inextricably part of multiple webs, tugging and being tugged.

It's stating the obvious, but the most visible relationship of the pinyon-juniper woodland is that between pinyons and junipers. The two trees do well in similar conditions. However, juniper has the ability to get water out of drier soils than pinyon can (because juniper can resist cavitation, as we have seen). This means that junipers can keep conducting photosynthesis, growing and reproducing at times when pinyon must shut down in order to keep from losing water through leaf pores. Junipers also tend to have more root mass than pinyons, giving them another advantage in dry country. However, at higher elevations, where water is more available, pinyons photosynthesize more efficiently than junipers and so can grow more quickly and get the advantage. These companion trees may compete, but one way they help each other is by acting as "nurse trees" for each other's seedlings. Both juniper and pinyon seedlings benefit from the protection of shade, at least when they are young.[5]

As for other relationships, even a mere list of the organisms within the PJ forests gives an inkling of the richness of the ecological web. Birds, for instance. In the pinyon-juniper, birds find food—berries, nuts, sap, and insects—and places to perch, sing, and nest.[6] Some birds live in the PJ permanently; some come for the summer or winter; some just dip in and out occasionally. Whatever their role, the following birds have been noted in various PJ forests of different types and at different locales:

- turkey vulture
- bald eagle
- northern harrier
- Swainson's hawk
- red-tailed hawk
- ferruginous hawk
- rough-legged hawk
- golden eagle
- American kestrel
- peregrine falcon
- prairie falcon

- chukar
- greater sage-grouse
- Gunnison sage-grouse
- mourning dove
- great horned owl
- northern pygmy-owl
- long-eared owl
- northern saw-whet owl
- lesser nighthawk
- common nighthawk
- common poorwill
- white-throated swift
- black-chinned hummingbird
- broad-tailed hummingbird
- rufous hummingbird
- downy woodpecker
- hairy woodpecker
- northern flicker
- gray flycatcher
- cordilleran flycatcher
- Say's phoebe
- ash-throated flycatcher
- Cassin's kingbird
- western kingbird
- loggerhead shrike
- northern shrike
- gray vireo
- Cassin's vireo
- pinyon jay
- Woodhouse's scrub-jay
- Clark's nutcracker
- black-billed magpie
- American crow
- common raven

Cassin's vireo (formerly classified as solitary vireo).

Pinyon jay, which has a limited range in the western and southwestern United States. The bird is currently under threat because of the loss of pinyon-juniper woodland habitat, where it gathers for feeding in massive flocks.

- violet-green swallow
- mountain chickadee
- juniper titmouse
- bushtit
- white-breasted nuthatch
- Pacific wren
- blue-gray gnatcatcher
- western bluebird
- Townsend's solitaire
- European starling
- cedar waxwing
- black-throated gray warbler
- spotted towhee
- chipping sparrow
- Brewer's sparrow
- lark sparrow
- black-throated sparrow
- Scott's oriole
- house finch
- common redpoll
- pine siskin[7]

Long-eared owl.

One resource manager spoke of PJ woodlands as being "rather sterile" in terms of energy flow, nutrient cycling, and diversity.[8] Unquestionably, many PJ woodlands have only sparse vegetation among overcrowded trees. But this generalization seems like a half-empty view of the glass. There are many ways to look at a glass, and in fact more than one kind of glass. In 1993, Bruce Koyiyumptewa, district silviculturist for the Coconino National Forest in Arizona, gave a picture of anything but sterility as he addressed the relationships the Hopi people have to the pinyon-juniper woodland. He said, "Hopi thinking is this: All Things Are Connected, the Earth, Plants, Animals, the Sun, Moon, and Stars, and Mankind." The interconnected plants and animals of the pinyon-juniper ecosystem are important to Hopi ceremonies, which are about having a "fulfilled, successful, and balanced life. Here the word 'successful' does not mean to be affluent in money or material things but simply to be in harmony with Oneself, with the Mother Earth, Plants and Animals and to have a happy, fulfilled, balanced life." He spoke of one young man who had taken several trips to the PJ woodlands with his godfather looking in vain for the sandgrass, alder, larkspur, buttercup, and other plants needed for the Soyal Ceremony. Nor could they find the feathers of hummingbirds, yellow warblers, American kestrels, and other birds needed for prayer feathers. This was so because for years the Indian agency had chained pinyon-juniper woodland to provide forage for grazing, planting nonnative plants to replace the natural, diverse vegetation. In addition, "wanna-be Indians" and herbal vendors competed for the remaining ceremonial plants. Without access to these plants and birds, how, in the Hopi view, could the people "start on their journey, this Road of Life to be in harmony"?[9]

*

In Lebanon, a juniper called *Lezzeb*, known scientifically as *Juniperus excelsa*, grows high on limestone cliffs. The Lebanese don't worry about Lezzeb spreading. They worry about it disappearing because of agriculture, woodcutting, and grazing. "Human beings everywhere pose a danger to the environment if they don't have awareness," says Mohammed Taleb, a man who has been trying to create a reserve to protect these junipers. These junipers keep the soil from eroding, and they support other plants, animals, and birds. In turn, thrushes support the junipers by spreading the seeds. Unfortunately, climate change and hunters have reduced the numbers of thrushes in Lebanon's mountains. "They are all related in one ecosystem," said Bouchra Douaihy, a scientist studying the Lezzeb. "Any threat to one can affect the others."[10]

Strange to say, people in the American West also once actively worried about the loss of junipers. During the first half of the twentieth century, people feared that the cumulative decades of cutting for mines, smelters, fences, and firewood would decimate juniper populations beyond recovery. But then values shifted. People moving to cities didn't need firewood or fences. A nation that during World War II had been on rations for gasoline, sugar, tires, silk, and red meat hungrily returned to its consuming ways. People wanted beef! So land-grant colleges trained numerous range scientists in techniques of improving grazing lands. The large expanses of pinyon-juniper woodlands beckoned.

"Most range managers recognize [land occupied by] pinyon-juniper as having the greatest potential for red meat production west of the 100th meridian," said one of those managers.[11] The country had a lot of military equipment left over from the war, and fuel was finally available and cheap. So land managers tore out almost one-third of a million acres of trees in Utah and Nevada[12] and put in crested wheatgrass and Russian wild rye, nonnatives that made for great spring forage. With more grasses growing, the government granted more grazing permits. Ranchers were happy. Almost everybody was happy, for a time.[13]

By the 1970s, though, environmental groups had begun to take notice. They noticed these "treatment" methods:

- Chaining: A length of anchor chain (forty-five to ninety pounds per link) is stretched between two crawler tractors; the tractors then drag the chain through a woodland. The idea is to rip out the trees and break up the soil for seeding. Between 1960 and 1972, 514,000 acres of pinyon-juniper woodlands fell to chaining in eight western states, half of those projects in Utah.
- Cabling: The same process, using a cable instead of a chain.
- Windrowing: A bulldozer pushes the trees into piles or windrows to be burned.
- Crushing: Heavy steel drums (ten feet long, four feet in diameter) with six-inch-deep blades replace the wheels on a huge machine, which lumbers through the woodlands and pushes down trees.
- Burning: Intentional fires kill the trees and clean the area for reseeding. If a fire gets too hot, however, it can sterilize the soil and kill the native seeds within it.[14] Clearing junipers with fire generally worked well during the early years, until cheatgrass invasion became common. This weed now quickly sprouts up in burned, bare ground. Besides pushing out native vegetation, cheatgrass itself is very flammable and spreads fire quickly.[15]
- Poisoning: Effective, but as one commenter pointed out, "When you kill these trees with a herbicide you've really created a greater problem than you had to start with because you've created a landscape that is unacceptable to the public. We've had that

> experience. When we get all done killing the tree with the herbicide then we've got the expense of going in and taking that thing out which is more expensive after it's dead than when it's alive."[16]

The results of all of these: big, stark scars of cleared land. The sight shocked people who loved the trees.[17] In 1974, Jennie Nylund, a high school biology teacher in Nucla, Colorado, wrote an emotional letter to the *Denver Post* about a proposed chaining nearby, in which she said, "We're going to lose the only beauty we have in this little valley."[18]

*

The animals that live in relationship with pinyon-juniper ecosystems may or may not see "beauty," but they do have to survive. Several mammals use the pinyon-juniper forests, finding hollow trunks, shade, protection from cold, and food—foliage, berries, or other animals.[19] Mammals that spend time or live in PJ woodlands include mule deer, elk, antelope, and bighorn sheep. Mexican, Stephens', white-throated, and bushy-tailed woodrats. Pinyon mice. Deer mice. Great Basin pocket mice, western harvest mice, northern grasshopper mice, canyon mice. Long-tailed and sagebrush voles. Least chipmunks, cliff chipmunks, Colorado chipmunks, rock squirrels, and white-tailed antelope ground squirrels. Also chisel-toothed kangaroo rats, mountain cottontails, porcupines and black-tailed jackrabbits. Badgers and weasels sometimes.

Stephens' woodrat eats practically nothing but the foliage of juniper, mainly one-seed juniper. Because the foliage is so poor in nutrition, the woodrat limps along in its life history; females reproduce late in life and may produce just one baby. Young rats grow slowly, and their mothers often die before they can produce siblings.

And then there are animals that eat animals. Mountain lions, which may kill two deer a week. Coyotes, which live mostly on rodents but also eat tremendous numbers of juniper berries and defecate the seeds. Bobcats, which eat rodents, birds, and fawns. Reptiles, including collared, brown-shouldered, and whiptail lizards and the Great Basin rattlesnakes and red racer snakes. Finally, there are invertebrates: ants, termites, cicadas, scorpions, centipedes, tarantulas, cone moths and cone beetles, and grasshoppers.[20]

Mountain cottontail.

*

PJ removal projects during the 1950s through the early 1970s cleared more space for sheep and cows, affecting all these organisms and their relationships. Over time, some observers began to question the assumption that tree removal and reseeding with grass was an unmitigated good. Game managers began to see some relationships between tree removal and decreased deer numbers, since deer, which don't eat much grass, were getting less feed and less habitat.[21] Watershed specialists realized that the relationship between treatment and decreased erosion wasn't that clear-cut. Range managers realized that in some cases they couldn't prevent new relationships, as when, for instance, cheatgrass and other nonnative plants quickly moved onto treated areas and outcompeted the desired grasses. People who analyzed the cost-benefit ratio of treatment found that the costs often outweighed the benefits, especially when gasoline prices soared after the Arab oil embargo of the early 1970s. Only in the best of circumstances—where the best forage could be grown or the trees could be put to some economic use such as particleboard or posts—did the projects make economic sense.[22] People began to recognize, too, that the removed trees would probably just come back, and they would have to be torn out again after twenty or thirty years.[23] Obviously, not all "treatments" had worked as hoped (an ecologist described one project, for instance, as providing "a useful lesson in how to destroy an aesthetic woodland resource by careless site and treatment selection").[24]

So, attitudes were evolving. One factor in all this was the National Environmental Policy Act (NEPA), which became law in 1970. Under NEPA, agencies had to write an Environmental Assessment (EA) and, potentially, a much more detailed Environmental Impact Statement (EIS) before doing a removal project on federal lands. Environmentalists quickly learned how to use these tools. Because of the stringency of the EA and EIS processes, because of threatened and real lawsuits (for instance, in 1975 the Natural Resources Defense Council sued the Bureau of Land Management over chaining), and because of disappointing cost-benefit ratios, the BLM essentially stopped chaining juniper in the 1970s and didn't begin again until the 1990s.[25]

Ranchers didn't appreciate this state of affairs, but the state of Utah hastened to fill the gap. The state gave interest-free loans so ranchers could clear trees from private and state lands. One cattle company accordingly took out thousands of acres of dense juniper forests in the Needle Range on the Utah-Nevada border. Chaining the trees preserved the ranchers' own economic and personal relationship to the land. A spokesperson for the company claimed that if they couldn't take out junipers, "the range wouldn't support the eight families that make up the company.... Our kids would have to move away." The cattlemen also said that the juniper removal enabled cows to produce more and bigger calves and caused a number of springs to start running. One stream "used to dry up by July.... Now there is water all year."[26]

The relationship between juniper removal and increased stream flows likely depends on the site and the trees. In the case of western junipers (*J. occidentalis*), as few as nine trees per acre can use all the water from thirteen inches of yearly precipitation. At Camp Creek in Oregon, the removal of all western junipers less than 140 years old increased the late-season flow of a spring by 225 percent, increased the groundwater, and increased the soil moisture available to other plants.[27]

In other situations, removal may not have such a dramatic effect on water flow. A couple of speakers at a pinyon-juniper conference in Logan in 1975 spoke about this. After one presentation, a questioner from the audience said,

"Am I understanding you right . . . when you say we are not actually improving the watershed when we convert from a pinyon-juniper to a good grass-browse-forb ground cover?"

"That's the story at the present time."

"I guess I've been brainwashed all these years."

"Don't feel like the Lone Ranger: there are still lots of people that feel that way."[28]

Over the years, continuing studies on PJ removal support the common-sense notion that different actions in different circumstances will produce different outcomes.

*

"One of the perils of being a scientist is that you never see things in black and white. There's always more going on," says Mark Brunson. "The connections and intricacies are what make life sciences so fascinating."[29] But land managers can't simply intellectually enjoy those intricacies and contradictions. They must grapple with the inevitably incomplete information they have, and with politics, pressure from stakeholders, and their own biases and goals. Then they must make real-life decisions. Almost certainly, any decision will raise somebody's hackles. In 1991 the U.S. Soil Conservation Service announced plans to chain and reseed 8,765 acres in the Muddy Creek–Orderville watershed (north of Kanab, Utah). This half-million-dollar project, the Soil Conservation Service said, would reduce erosion by 39,528 tons yearly. The reduced erosion would then reduce salinity in the Colorado River. However, Ken Rait of the Southern Utah Wilderness Alliance (SUWA) called the agency's assumptions "absurd and dubious." Even if the project really did prevent that much erosion, it could change the salinity in the Colorado River by only 0.0002 percent, he said. A Sierra Club spokesperson added, "What they really want to do is create forage for livestock." Indeed, the project would provide more than fourteen square miles of new rangeland, so that a whole lot more livestock could graze there—another 4,209 animal unit months, or AUMs, per year. (One AUM equals the amount a cow plus calf, one bull, or five sheep will graze in a month.)[30] A number of citizens also opposed the project, and even a BLM employee in Arizona commented, "Reducing soil erosion and salinity from watersheds is the oft-stated major benefit from chaining and seeding of pinyon-juniper woodlands. But research does not show such significant slowing or reducing of erosion by chaining and seeding. The real goal is to increase forage production, and it should not be masked behind imaginary erosion control benefits."[31] Three years later, the BLM withdrew from the project, citing lack of personnel and money; Ken Rait believed that the "BLM backed out because it is beginning to recognize that chaining is an economically and technologically indefensible range tool. And I believe public pressure hastened this education process."[32]

Despite decades of grazing focus, today cows are no longer the top priority for most people at the BLM and Forest Service, according to Mark Brunson. True, livestock grazing is still part of the land management mix, he says, but agencies remove PJ mainly to reduce fire risk and erosion, get rid of cheatgrass, and improve sage grouse habitat.[33] On the other hand, Doug Page, a retired forester, has observed a lot of varying opinions within the BLM, ranging from "get rid of all the junipers" to "keep them all," and everything in between. There's usually not a consensus within agencies on how to manage any piece of land, he says. The BLM's forest, range, and recreation specialists may be at odds, disagreeing about how much juniper has spread since settlement—not to mention about how much should remain on the landscape now.[34] Project decisions and justifications, then, might well be shaped by competing internal opinions and endeavors—and in the end, that may be one reason why environmentalists' traditional perceptions about agencies continue. "The bottom line is, I would like the agencies to be honest about the reasons for the treatments," said Allison Jones of Wild Utah. "If the reason for the treatment is to create more forage for cows, then say so. Don't hide behind statements about improving watershed, restoring the national fire cycle, or creating more brood-rearing habitat for sage grouse. They are shrouding their reasons because they're being scrutinized by the environmental community a lot more."[35]

That scrutiny is a good thing, says Fee Busby, professor of wildland resources at Utah State University; there's a relationship between environmentalist pressure and the better treatment projects taking place now. Today's projects really do take into account things like wildlife and hydrology instead of just cattle, he says. For instance, instead of planting only nonnative grasses for the benefit of cattle, agencies now include native grasses and plants. Busby says, "You have to credit the environmental community for those changes. Environmental communities have been an amazing positive force."[36]

The environmental groups themselves probably wish they could be a larger force, able to actually stop projects they see as harmful. Neal Clark of SUWA says that outside of oil and gas development, pinyon-juniper treatment projects pose probably the biggest threat to wilderness-quality lands in Utah. The BLM often tries to reassure wilderness advocates that in the long term the treatment projects will enhance wilderness characteristics, he says. But later, when the BLM evaluates treated land for wilderness designation, it can reject the designation by saying that the project has destroyed wilderness characteristics. "They're using both sides of the same card," says Clark.[37]

*

On a July day in 2010, I went out to Rush Valley with Keith Olive, fuels specialist with the Salt Lake office of the BLM, to see a juniper treatment project in an area called Big Hollow, on the south end of the Stansbury Mountains near Johnson Pass. Driving out west from Salt Lake City, Olive told me he'd graduated with a degree in American history, moved to Utah in 2000, and then become intrigued by all the wildfires in the West. He ended up with the BLM fuels crew. Besides fighting fires, this crew assists with fuel reduction projects—juniper treatment, in other words. "At first I didn't understand why we were killing all these juniper," he said. "Now I see the advantages.... The Rocky Mountain Elk Foundation [a hunting organization] loves it. Out on the Deep Creek Mountains there's been a herd of sixty elk running around in the project area. I think it's a huge benefit." As we drove across Rush Valley, I could see several hillsides that had already been shaved of trees—khaki patches among the dark cedars. Olive, originally from Georgia, couldn't believe how hot, dry, and pale the West Desert was when he first saw it. Now, he said, "I like the desert. I think it has its own beauty."[38]

People began to live at Big Hollow centuries ago, and archaeological sites suggest the relationships humans have had with the juniper ecosystem. PJ forests contain a lot of archaeological sites, more than there are in higher forests or on sagebrush benches. Archaeologists regularly find fifty sites per square mile of PJ, and in the most ideal areas for habitation they might find one hundred.

In particular, ancient people liked to winter among the trees.[39] "I love it that prehistoric people were all about the juniper," says archaeologist Lori Hunsaker. "The PJ was a much warmer place to spend the winter. They got it."[40] Before an agency can remove the junipers in a place, by law it must have archaeologists survey the area to locate any sites and recommend appropriate action.

There are humans living in the juniper at Big Hollow right now, and this project was motivated in large part by their presence. A remote subdivision with several homes has gone up within the trees here. Because of this subdivision, the BLM decided to do a Wildland-Urban Interface project. The project was designed to create greater water flow in the spring that serves the homes as well as to lessen the chance of wildfire, or at least the chance of a monster wildfire. Olive explained that it's easier to fight fires when they're burning on the ground, not in tree canopies. He described a 2007 fire in northern Utah that was burning hot and out of control from tree to tree until it came to a place where the juniper had been cleared. There it dropped to the ground, and the firefighters could get to it and put it out.

According to professor Fee Busby, during the last couple of decades the fear of wildfire has been the biggest driver for juniper removal. By the 1990s, wildfires had become almost epidemic during the summer months. Those big fires "scared people to death," says Busby. Managers realized they needed to reduce the fuel, especially around homes and buildings. By then, a new method of taking out trees had emerged. A mulching machine called a Bull Hog had been used in forestry to grind down stumps, and when brought into the PJ, it took out trees fast.[41]

Three of these mulchers were now at work finishing up the 777 acres of this phase of the Big Hollow project. The BLM had told the contractors to avoid old-growth junipers and understory plants like mountain mahogany, bitterbrush, and serviceberry. It had also instructed the contractors to leave trees in drainages to provide corridors for wildlife and to "feather" the edges of the treatment. But Olive pointed out some large bare parcels where one contractor had messed up and clear-cut everything anyway.

We passed an area of naked gray and black trees, a burn from two years earlier. In another area, trees had been cut, piled up, and covered with plastic to keep them dry. In the fall, after a rainstorm, the BLM would burn these piles.

Bull Hog masticating a juniper in Dammeron Valley, Utah, in a 2009 project conducted primarily for wildfire protection and secondarily for habitat benefits for wintering mule deer. Courtesy Bureau of Land Management, Utah.

We saw a deer picking her way through a dense grove of young junipers. We passed an area that had been treated and seeded in 2007. Some of the old slash still lay strewn about, but the seeded plants—a mix of native and nonnative—had sprouted: crested wheatgrass, bluebunch wheatgrass, Indian ricegrass, flax, and yellow clover. The new meadow looked and felt fresh and pleasant. A few big trees still dotted the slope.

I could hear the roar of the machine as we approached it. The slope had been cleared of countless trees, and mounds of shredded wood lay among the sagebrush. The smell of cedar filled the air. A yellow excavator was clanking, grinding, whining. Its arm swung up and above a medium-sized juniper, the mulching roller on the end of the arm rapidly spinning, and then it descended straight down the trunk. In three seconds it had mowed the tree to the ground. It just obliterated it. Bits of wood and leaves flew up. The excavator arm swept from side to side to clean up the trunk and lower branches and then moved to the next tree. Zzzzrp, and that tree was gone.

Up in the cab controlling this beast—for it really did seem like some prehistoric animal snorting and annihilating vegetation—sat a teenage girl, daughter of a subcontractor. She and her brother were out here working this project. She knew what she was doing. She could fix these machines, she could drive them, and she could annihilate trees. The machine growled, nosed the ground, and maneuvered forward to the next tree, stray branches in its mouth. Zzzzrp. Tree gone. The mulcher kicked up a spray of wood and bark. I felt a little heartsick. Maybe this was the greater good—but a tree snuffed out, like that? I thought of my own garden and pulling weeds. Was it just that—weed pulling? Tree after tree vanished.

Olive told me it costs $250 per acre to treat a site this way. You could chain it for a lot less. With a mulcher, though, you can clear a thousand acres in a couple of weeks and be done with it; no need to deal with acres of uprooted trees.[42]

*

Though junipers seem strong, clearly there is always something stronger. It's not always something big and heavy like a Bull Hog. Several relationships with small things can also hurt, kill, or limit the trees: rust fungus, leaf rust, wood rot, blight. Flatheaded borers attack the wood. Roundheaded borers attack twigs and limbs. Longhorned beetles girdle limbs and twigs. Caterpillars eat the berries and into the seeds. Mistletoe—pale, yellowish, sickly green, with white berries—parasitizes junipers and weakens them, but it also attracts birds, which eat both mistletoe and juniper berries and subsequently "plant" the seeds. For its own part, juniper limits other plants, like grasses, by covering the ground with litter and sending out aggressive roots.[43]

Relationships go both ways—that is, they go all ways. Every organism affects the environment somehow, capturing nutrients, water, light, and heat, returning through excrement or its decaying body some of what it has taken. And organisms provide something other organisms need, like shade for seedlings, nesting areas, and food. Everything is connected.

The 1958 essay "I, Pencil" lists the truly mind-blowing number of processes, products, and people that go into making a simple lead pencil. The essay argues the classic conservative stance: governments should let the "Invisible Hand" of economic self-interest do its thing with a minimum of interference.[44] What the essay doesn't explore is the ripple effects of everyone seeking their own interest. What happened to the ecosystems affected by the logging; the mining for iron, zinc, copper, and graphite; the steel fabrication; the milling; the making of the pencil "lead"; and the final manufacturing of the pencil? What pollution came from all these processes, and what did the pollution do to lives and health? What was the quality of the health and safety and wages of miners, agricultural workers, factory and mill workers, and railroad and dam builders around the world during the making of the pencil? On the other hand, how many people have benefited from having cheap pencils available? How have people grown in knowledge and skills? How have businesses prospered? Everything is connected.

We tend to focus on only those strands of the web that mean the most to us, which can lead to polarization. In 2007 the Utah Division of Wildlife Resources (DWR) announced the chaining of one thousand acres of PJ on Tabby Mountain in the Uinta Basin of eastern Utah. The project, the DWR said, would create a natural mosaic of grasslands and trees, improving the watershed and increasing deer and elk populations (thus furthering the DWR's main mission). The oil and gas giant Bill Barrett Corporation chipped in some money as mitigation for habitat damage it had done elsewhere. The Forest Service contributed money, as did the Rocky Mountain Elk Foundation. All of these entities had their own interests. But an online commentator had different interests. She collected and sold pine nuts for a living, and she wasn't buying into the justifications for the project. A project like this "is why there are no pine nuts," she argued.

> Mule deer, bear, elk all eat pinyon pine nuts. Basically, this project is not about wild life. It is about cattle. Taking away a primary food source need[ed] for winter survival of the wild life will not benefit them. Not to mention that wild American pinyon pine nuts go for up to $30.00 this year. Many, many people benefit as well as the wild life from the pinyon forest."[45]

In another polarized situation, Vickie Tyler of the BLM spoke of the significant time she'd spent trying to communicate the BLM's point of view to a representative of SUWA. "It's not like we sit in our office and go 'Hmm—I'll go clear this piece of land.' A lot of biology goes into it," she said. "I have a degree in public affairs and biology so I can help people understand what we do, but it's really hard to get it across. You hope when you spend time with people it will make a difference." However, in this case, SUWA responded with "a forty-page letter telling us what to do."[46]

This kind of conflict has been going on for a long time. In 1975, one of the speakers at a pinyon-juniper conference lashed out at the whole endeavor of PJ management. He accused the agencies, range managers, and scientists of manipulating nature for the benefit of special interests and self-interest. He said agencies were playing the role of conqueror instead of working within the ecosystems. He claimed they were not using good science; they were basing the "massive destruction of ecosystems" on mere "guesswork and wishful thinking."

A member of the audience objected:

> I always enjoy the environmentalist's position where he can come in off the cuff, level the blast, and leave. As you indicated, you have been here for only part of the symposium.... If you had been here for the total symposium, I think you would have seen some very conscientious efforts by various people from throughout the entire southwestern United States to look at some of the main problems and come up with non-biased answers.... We don't have a good way for putting watershed, big game, range, aesthetics, and whatever you'd like to throw into it, together into one formula and come out with an answer. We haven't reached that point yet. It is extremely difficult to weigh these things and come down to an answer, and it depends on many many factors.[47]

The environmentalist spoke true: too many times, land management decisions have been made more for special interests than for the health of the ecosystem. Many times, decisions have been based on incorrect assumptions and hypotheses, or maybe even at times on guesswork and wishful thinking. But the audience member also spoke true: a one-dimensional criticism is unfair.

Although politics is still a major force in western land policy, increasing numbers of biologists, range scientists, foresters, ecologists, and managers were even then (and are still) investigating the multiple relationships affecting pinyon-juniper. They are weighing these relationships as they seek to restore degraded landscapes. Some are researching intact, old-growth ecosystems and speaking out to say that we should avoid damaging these ecosystem webs. They are saying that we can't always restore what we have broken.

10

Restoration

What is restoration? That is, when people speak of restoring an ecosystem, what does it mean? Does it mean the same thing to different people?

To explore this, in 2010 I asked representatives of the Bureau of Land Management and of the Southern Utah Wilderness Alliance if they could enlighten me about juniper removal projects. They both mentioned a proposed project on Upper Kanab Creek in Kane County, Utah. The project lay partly within the Grand Staircase–Escalante National Monument and also two proposed wilderness areas. Both women had plenty to say.

The spokesperson for the BLM, Vickie Tyler, explained that juniper is a shrub that must be managed. Removing the pinyon and juniper and thinning the sagebrush would improve habitat for sage grouse. "We're doing this on the Parker Range [in south-central Utah], and now there is three times more sage grouse," she said. The project would also protect the vulnerable black sagebrush from annihilation by fire, which destroys it beyond recovery: "When black sagebrush is gone, it's gone." The treatment would work much like a fire, which burns "fingers" of clearings into the trees. The mulch created by the Bull Hogs would discourage cheatgrass, but not desirables. "With the Bull Hog treatment, the land looks very stark at first," she said, "but we're seeing changes come back in a short period of time."[1]

Tiffany Bartz, attorney for SUWA, had a different perspective. "The Timber Mountain area northeast of Kanab has a pretty intact and diverse ecosystem of 130,000 acres," she told me. "The BLM is proposing deforestation on 51,600 acres." A SUWA fact sheet on this "vegetation devastation" proposal explained that the plan would take out 90 to 100 percent of pinyon and juniper in some areas and also kill sagebrush. The fact sheet charged that the BLM had not documented whether "treated" areas actually thrive in the long term, or whether it would be more effective to remove cows instead of trees. Nor had the BLM included a monitoring plan to analyze the outcomes. The fact sheet recognized that vegetation manipulation may be appropriate sometimes, but not in the Upper Kanab Creek area.[2]

Bartz urged me to visit the area to see the situation for myself. A couple of months later, I did. A dirt road goes east out of the little town of Glendale, Utah, up onto the Glendale Bench, a huge, wide, rolling country of thick PJ woodlands and sagebrush flats. The Pink Cliffs, one of the "stairs" of the Grand Staircase, rise up along the north side of the bench.

Up on the bench, someone had cleared trees from a piece of ground. A few big junipers had been left growing along the road, apparently to shield the view of the treatment area, but they couldn't really disguise it, especially since a mud-coated D6D Cat with rusted treads stood nearby. A five-gallon bucket of 15W-40 oil and a gas can sat on the seat, perhaps ready for the next pass. I pushed past the "shielding" trees and stood looking at mayhem. Heavy equipment tracks had dug into the mud, now bare of live vegetation except a sagebrush here and there. The corpses of trees, juniper and pinyon, lay bulldozed into rows. Broken limbs and splintered stumps were all piled in jumbles.

In another place and another culture, Roy Haiyupis of the Nuu-chah-nulth nation on the west coast of Canada had this to say about cutting a tree: "Talk to it like a person. Explain to the tree the purpose, why you want to use it—for the people at home and so on. It may seem like you're praying to the tree, but you're praying to the Creator."[3] Standing on the rutted ground, I couldn't imagine that a lot of praying had gone on here before the D6D started rolling. Obviously, though, somebody cared a lot about this land and what grew here, enough to spend money and put a lot of effort into changing the vegetation. The western sun threw my shadow across the scarred land as I walked over to the dead trees. Their smaller branches still held some life, leaves, and suppleness. Many of the trees had been magnificent old ones, tall and wide. The brilliant pink cliffs to the northeast glowed in the late sun, flanked by dark PJ forests.

It was dead quiet up there. Wide clouds floated above, their shadows moving across the sagebrush and woodland. A barbed-wire fence ran along the edge of the cleared area. On the other side of the fence, the vegetation remained untouched. I climbed through the wire and came upon a big old tree right next to the clearing. Biologists are finding that plants can communicate in certain ways, and I wondered whether this tree had somehow felt the destruction.[4]

I walked through the woodland—what the cleared area would have looked like before removal. The trees here did not seem particularly crammed together. Big, medium, and small ones stood in the grove, with sagebrush and grass tufts scattered between them. Also tiny yellow flowers with lacy leaves, prickly pear, bitterbrush, and a biotic soil crust. One ancient juniper stood maybe thirty-five feet tall. This looked like a diverse and ecologically intact place to my nonexpert eyes. As I walked farther into the woods, no kidding, I felt a peaceful sweet atmosphere. A breeze started somewhere, making the trees sing, coming closer until the trees above me swayed.

Back at the cleared area, I looked over the bare soil and tried to imagine the eventual outcome: What new plants might be growing here next spring and the spring after that, and how long might it take for this area to look "natural" again? Some studies have found that chaining increases grass cover at first, but over time, when combined with grazing, it results in more bare soil and a lower amount of biological soil crust than in untreated areas. A treated area can also provide perfect conditions for trees to return. Mature seeds from the downed trees and in the soil can germinate and grow, helped along if slash piles of dead trees remain to provide a shaded, moister area for germination. On chained sites, junipers tend to come back more quickly than pinyons do. So unless you re-treat a site every couple of decades, the trees will quickly recolonize. The catch-22 is that if you do re-treat it often, the junipers will still have the advantage and become dominant. At least, that's the story in some places; there is still a lot to learn about what pinyon and juniper removal does to vegetation structure and soil.[5]

In the end, I wondered, what would be gained and what would be lost through a tree removal project in this area?

*

We have undeniably degraded much of the land in the West—"we" meaning the collective culture as well as individuals. If we now try to make some amends by "restoring" some of the land, how do we do that? One esoteric definition of restoration is "reinstatement... to a prelapsarian state of innocence; salvation, redemption."[6] If only there had ever been such an innocent state, a single pure moment in time and space! But there is no such thing. Even though some have argued that we should return the land to what it was like before Anglo settlement, there was no static, ideal condition in the past, and even if there was, we couldn't re-create it. In fact, we can't even say for sure what the land was like before settlement, or before humans arrived. That's because everything changes constantly. So how can anyone say definitely what the ideal natural condition of a landscape is, was, or should be?

What we can say is that we want to restore a landscape to health and resilience, but first we must define what "healthy" looks like—or, more precisely, we must decide which vision of "healthy" to pursue among the definitions various stakeholders put forth. Juniper removal will always benefit some species and hurt others. And although juniper removal may in some cases contribute to a healthy though changed local ecosystem, it also affects larger systems; for one thing, it eliminates the "huge service" the trees give to the world by removing carbon dioxide from the atmosphere and storing it, in the words of Jayne Belnap. "We find ourselves in an interesting conundrum of having to define what we want," she says. "That will become our baseline, because we'll never get a piece of land back to 'where it was.' We now have many competing uses for the same piece of landscape, with conflicting values. So it's really sticky.... Any time there's a 'treatment' there's a conflict."[7] Imagine the different values of a backpacker, an off-road vehicle enthusiast, a rancher, a hunter, a forester, a range scientist, a politician, an environmental group. When conflicts arise from people's deepest values, finding common ground is hard. It's similar to the prolife/pro-choice debate. You can't resolve that issue by negotiating an acceptable number of abortions. For instance, if a group believes that livestock absolutely don't belong on the land, they will not be willing to negotiate an acceptable number of cows.[8]

In the twenty-first century, one value driving juniper removal is wildlife habitat, and much of the funding in Utah comes from the Division of Wildlife

Resources. "I've seen projects where the BLM signed on to projects that DWR is really doing," says Neal Clark of SUWA. "It's about new habitat and new revenue from hunters." As of 2014, the DWR had budgeted a half million dollars for a removal project at Beef Basin/Dark Canyon, in the Four Corners region, a highly controversial project. In the Environmental Assessment for the proposed "Beef Basin/Dark Canyon Plateau Sagebrush Restoration," the BLM defines restoration: "'Restoration' of the project area does not necessarily imply an objective of returning an ecosystem to a condition that may have existed at a point in history, but rather the restoration of functional processes required to sustain resource values."[9] These "resource values" could include grazing land, big game habitat, water, and wilderness. But is a healthy ecosystem the same thing as an ecosystem manipulated to restore "functional processes required to sustain resource values"?

Perhaps not, if the treatments need to be done again in twenty or thirty years. It may seem obvious, but junipers will not grow in places that don't work for them, even when those places have been grazed heavily.[10] Animals and birds spread juniper trees. So if the trees start growing where they haven't grown before, there are two possible reasons: the animals and birds are putting the seeds in places where they did not put them before; or the site conditions have changed, and the seeds being deposited there are now able to succeed.[11] Either way, you have to figure out what caused the changes and address those if you don't want to be merely "landscape gardening," as SUWA's Neal Clark calls it. "Are we going to fight 'encroachment' indefinitely?" he asks. "Are we on the hook forever to keep treating it. . . . If the goal is to bring these ecosystems back into some type of natural balance, then the landscape gardening approach currently (and historically) applied by BLM is an abject failure."[12]

Psalm 23 talks of a Lord who "restores my soul." Even the sound of this phrase evokes a more holistic view of restoration, something beyond "functional processes" and "resource values." Even an exhaustive Environmental Assessment can't completely account for the soul of an ecosystem, which we might define as all those interconnections that animate a community, that sustain it, that make it resilient in times of change. Where we have damaged or destroyed these interconnections, thoughtful restoration may be needed.

*

About forty-five years ago, a wealthy guy named David Bamberger bought "the most worthless piece of land in Blanco County," Texas, according to a Soil Conservation Service staffer. This 5,500-acre abandoned ranch was eroded, gullied, dry, and overrun with Ashe juniper. Fewer than fifty species of birds lived on the land. The deer running around the place were scrawny. Bamberger took restoration on as a challenge, and he put employees to work with chain saws and bulldozers to take out thousands of junipers. He left some old-growth groves, planted three thousand new trees of many kinds, restored springs, and installed cisterns. In 1997, a visitor described this once-damaged land as "a Hill Country dream. Streams run year-round. Grasses are tall and lush. Diversity rules." The number of bird species had gone up to 148. Trophy deer attracted paying hunters.

Bamberger considered himself a steward of the land. He told a reporter, "Kids growing up in a world of McDonald's and graffiti are going to be making the decisions in the future. I want them to see what their world can be like. That's why I'm telling the private landowners they're holding on to a myth. It's not like it was one hundred years ago. We don't own this land. We're only stewards."[13] From the sound of it, he has restored the land's "soul," and that involved, partly, taking out junipers.

Landscape management has been around for a long time. In one instance we know of, the Timbisha Shoshones of Death Valley managed the vegetation of their own habitat, keeping pinyon and mesquite groves clean and free of litter and underbrush. They trimmed the lower branches of trees, removed dead branches, and raked away litter and duff. This cleanliness kept blowing sand from piling up and choking the mesquite, and more importantly it also kept wind from turning campfires into wildfires. The Timbisha also pinched back pinyon pine and leafy plants to encourage new growth, and they kept water holes cleaned of debris. They did this for practical and aesthetic reasons—in part so that the land would look cared for. As a result, many of the woodlands around the valley were open and free of underbrush. Today, having been left to natural processes, the groves look "unkempt with all of the dead wood and debris," and hummocks of sand have piled up against the trees and brush.[14]

Marshall Jack of the Owens Valley Paiute/Central Miwok/Washoe Tribes said his grandfather told him how tribes east of the Sierra Nevada managed

A juniper in the ecosystem of Vermilion Cliffs National Monument, Arizona.

the land they lived on. "They burned to increase foods such as wild onions, elderberries, and caterpillars, and to clear out the underbrush to bring in the new growth for the animals. . . . Crawley Lake used to have a lot of pinyon pine trees. They'd burn every three years to increase cone production the following year and to decrease the grasses and duff from the pines. It would keep them from losing pinyons from a really big fire."[15]

If humans have managed the landscape for hundreds or thousands of years, they likely affected the state of many pinyon-juniper woodlands as settlers found them. But again, the level of manipulation by aboriginal people depended on the geography, the vegetation, the landscape history, and the people's cultural needs and practices. This is not so different from today. Differences between habitats and their histories have led to different situations. In all three of the major pinyon-juniper types—persistent woodlands, PJ savannas, and wooded shrublands—trees have increased dramatically in some places. In the absence of wildfire, some of these places have benefited from a careful, well-planned thinning of young trees, along with changes to grazing regimes.

*

Clearly, though, an intact ecosystem does *not* need to be restored. If we manipulate such a landscape to meet some human need or wish, we certainly can't call that restoration. In particular, old-growth pinyon-juniper ecosystems are unique and irreplaceable. They are a key component in the West's biological mosaic. They are healthy. On the Kaibab Plateau in Arizona, for instance, undisturbed old-growth sites had more plant diversity than grazed sites, and much better cover by biological soil crusts (BSCs)—in fact, thirty-three times more BSCs than heavily grazed sites had.[16] An ancient forest like that cannot be improved by treatments. Historically, however, too many chaining projects have not distinguished between young woodlands and several-hundred-year-old woodlands. "We get very alarmed about recent invasion, but when it comes to clearing the woodland it's old woodland that we clear," one biologist noted in 1975.[17] Back then, some people actually supported removal of old trees. To their mind, "old, stagnating stands" needed severe thinning to remove "overmature and mature vegetation." This would maximize the woodlands' commercial potential to produce more firewood, fence posts, Christmas trees, pinyon nuts, essential oils, particleboard, veneer, pulp, turpentine, and resin.[18]

Yet even today the ecological values of old-growth pinyon-juniper "have been under-appreciated,"[19] and projects too often include old-growth removal. Just because pinyon and juniper are thickening in some places, "many managers throughout the West have tended to view the entire piñon-juniper vegetation type as degraded and in need of intensive restoration," wrote authors of a study of pinyon-juniper on Mesa Verde. The National Fire Plan, written after a devastating fire season in 2000, has encouraged major tree and shrub removal efforts in the West through thinning, clearing, and prescribed burning. However, the national policy driving this removal does so through generalized assumptions about fire in pinyon-juniper—assumptions that are not based on the latest science.[20] Even for fire mitigation purposes, old-growth PJ generally does not need to be restored or thinned. In fact, the old forests on Mesa Verde haven't

Rocky terrain, where fire can't easily spread.

burned much in the seven hundred years since the Ancestral Pueblo people abandoned their villages there. Even though lightning strikes the mesa many times each summer, fires don't spread; the rocky high-desert terrain keeps the flames contained. "Considering that these are some of the oldest forests in the southwest, and that they support a variety of old-growth fauna and flora," say the Mesa Verde authors, "their conservation value should be considered before subjecting them to what may be essentially irreversible changes."[21]

Those old-growth groves that have been removed can't return to that condition in our lifetime, to say the least. And besides, treatments may rearrange ecosystems so thoroughly that the ecological patterns and processes cannot be restored.[22] We can't assume that we can ever completely "fix" land that we have disturbed. Ethnobotanist Nancy Turner reminds us, "Even if Nature worked like a well-run machine—and we know that it is much more complex than this—we have been demolishing the machine without keeping track of the pieces, or in some cases without even keeping the pieces. Any attempt to rebuild it is bound to be defective."[23] Despite growing environmental awareness, the demolition of ecosystems continues around the world, including old-growth pinyon-juniper woodlands. Perhaps destroying old trees in the past was a mistake. But now we know better, and you can't call the removal of old-growth forests a mistake. It is a willful assault, a failure.

*

A healthy landscape has a good mix of native species, balanced exchanges of energy and nutrients, resilience to disturbances, resources for animals, and endurance—while evolving—over time. A monoculture of same-aged trees with little understory is not the healthiest community. Leaving overcrowded, same-aged junipers in place may cause increased erosion, a bigger risk of big fires, and the loss of sage grouse habitat.[24]

At the same time, there are pitfalls to removing them. If you do thin junipers:

- The process of thinning will destroy soil crusts, if they are present.[25]
- Flammable cheatgrass and other invasive weeds could move in to make up a large percentage of the posttreatment ground cover. The Mesa Verde researchers stated that extensive thinning and burning of PJ in that area could create an "ecological disaster." As flammable weeds move in, more frequent fires could destroy most of the old-growth forests relatively quickly. "Such a loss would be especially tragic if it resulted from well-intentioned but misguided efforts to 'restore' the piñon-juniper ecosystem."[26]
- Without a change in the basic cause of juniper spread or infill, the trees will come back, and the land will need re-treating.[27]
- Water flows might or might not improve.[28]
- Livestock forage might or might not improve.[29]
- Mulching with wood masticated by Bull Hogs might have unintended consequences. "We don't know anything about juniper mulch," Jayne Belnap says. "It doesn't decompose. Juniper wood can last hundreds of years. We have no idea what the mulch several inches deep spread across the landscape will mean for plants in the long term."[30]

Mill Creek area in Kane County, Utah, showing the area during treatment in 2008 and three years later. Courtesy Bureau of Land Management, Utah.

- Some species will benefit and others will not. "It would be unusual to not hurt anything when you take out juniper," says Allison Jones of Wild Utah. "You have to be clear on why you're doing the treatment. . . . What we need is full disclosure and a holistic approach—something we don't often see in the Environmental Assessments for these treatments."[31]

*

In 2000, San Juan County (Utah), the Charles Redd Foundation, and Utah State University Extension produced a book stating that Utah rangelands were in the best condition they had been in for one hundred years, and that they were continuing to improve. Starting in the 1940s and 1950s, the government reduced numbers of livestock on the rangelands of the West, shortened grazing seasons, and changed the distribution of livestock. The government also did some vegetation treatments.

Repeat photography in the book indeed shows healing. In the older photos, stream channels looked wide or gullied and sparsely vegetated. At several sites, only scattered vegetation grew. The recent photos show stabilized streambeds flanked by thick riparian vegetation—though much of it is tamarisk and Russian olive, undesirable exotics. On ground that cows had once stripped of vegetation, diverse healthy grasses, forbs, and shrubs now grow. Most of the photos actually don't show much change in the juniper—neither expansion nor removal. But in all, the photos make a hopeful case that land can heal and come into better balance if given a chance. The adjustments to grazing seem to have been key.[32]

Other places also show nature's resilience, when given a chance. Early in the twentieth century, Pine Valley in Utah's West Desert had been severely overgrazed. Cattle and multitudes of sheep had eaten and trampled a "severely destructive" swath through the valley. The valley suffered the usual consequences being played out across the West: decimated grasses and forbs and severe erosion. This particular Pine Valley straddles the Millard-Beaver county line and lies between the Needle and Wah Wah mountain ranges. Here in 1933, scientists did a detailed study of vegetation—what was left of it. In response to dismal land conditions like this across the West, Congress passed the Taylor Grazing

Act in 1934 "to stop injury to public grazing lands by preventing over-grazing and soil deterioration, to provide for orderly use, improvement and development, to stabilize the livestock industry dependent upon the public range." This act reduced the numbers of livestock allowed on public lands to more sustainable levels. In the 1950s, livestock on the land was reduced further and a band of feral horses was "removed."

When researchers resurveyed Pine Valley's vegetation in 1989, they found that the land had been slowly recovering over the decades. Perennial grasses and shrubs were coming back into dominance over trees. Among the trees, dominance was shifting from junipers to pinyon pines. Thus the valley, originally named after those pinyons, seemed to be welcoming the return of the plant community most suited to grow there.[33]

These sites had a chance to heal because of less grazing pressure. Today, after removing junipers and seeding, the BLM typically eliminates grazing for a time to let the grasses get established before allowing cows back on the land.[34] This practice is probably not going to change. But if grazing helped cause juniper spread in the first place, then how and how much the cows are grazed is important. If a landowner doesn't address the factors that caused changes on a site, a "restoration" effort will probably be ineffective or could even be harmful.[35] Whether you're talking about a human body or an ecosystem, treating only the symptoms can't create healing. Only the body or the land can heal itself, once you address the root cause of sickness.

*

However, climate change—one of the root causes of juniper expansion—adds another layer of complexity to the whole endeavor of restoration. As people work to restore damaged ecosystems, the world is crossing a threshold. Climate change has apparently encouraged the expansion of shrubs not just in the West, but on every continent except Antarctica. In fact, even if temperatures were not climbing, the increased carbon dioxide in the atmosphere has already changed the balance between herbaceous and woody plants in African savannas.[36] Being evergreen, junipers have a particular advantage; they can use the increased carbon dioxide year round.[37] The expansion of shrubs everywhere will cause a fundamental shift, affecting the animals that depend on grasslands, affecting the cultures that depend on the grasslands, affecting livelihoods. Herders will need to shift to browsers like camels and goats, and that will further affect ecosystems. Although land stewards can and do take out shrubs and plant grasses, if climate change is a major contributor to the spread of shrubs, are we just plugging a failing dike with a few fingers?[38]

"How smart can we be," some ecologists have asked, "and how much hubris is there in presuming that we can understand and predict ecological change?"[39] Clearly, we're not smart enough to know everything. Ongoing research, learning, and dialogue keep adding to the picture, even though it's hard if not impossible to know for sure how a restoration project will turn out as the climate continues to warm up. Still, after studying and analyzing specific sites, refusing to act from a one-size-fits-all mentality, and not caving in to political pressure by privileging one stakeholder over another, land managers need to make decisions. And they may decide to take action to sustain or restore the health of an ecosystem in their stewardship.

It may be naive to believe that ecosystem health should always be the top priority driving decisions of federal and state land agencies, but one can always dream. Even if decision making is too often politicized, we can't discount the many people who are working together to understand how to restore and sustain the health of the land. We also can't forget that others, deliberately or unknowingly, are undermining that health. But the reality is, most of us sit on the sidelines. Perhaps if this book, this saunter through junipers, has helped us sense more deeply the interconnections of the web we are part of, we might choose to get off the sidelines. There are so many things that need restoring and sustaining: a specific piece of land or woodland, our own relationship with nature, the wider human kinship with the living world. Junipers, as one of the living things that share this earth, can teach us something about all of these.

11

Kin

"Why not go into the forest for a time, literally?" Carl Jung advised a colleague. "Sometimes a tree tells you more than can be read in books."[1] If the book you're reading now has only deposited a little more knowledge into your mind, it's not worth much. After all, many of us already live in our heads too much of the time. But using that mind knowledge, perhaps both you and I can take a greater awareness of the many interconnections of nature into the woodlands—or into any natural place, for every place is full of stories. "On the ground," we can experience a web of being more complex and stunning than we could ever intellectually grasp.

Practically speaking, the earth's ecosystem webs are the basis of everything we experience and possess as organisms. Ecologists gesture toward these networks when they talk about ecosystem services:

- Supporting services: habitats, oxygen, soil, and the cycling of nutrients.
- Provisioning services: building materials, clothing, medicines, food, clean water, and everything else.
- Regulating services: carbon storage, air and water purification, shade, pollination, and erosion and flood control.
- Cultural services: spiritual renewal, recreation, education, and aesthetics.[2]

We have explored many of the provisioning services of pinyon-juniper ecosystems: wood for structures as varied as fences and cliff dwellings; fuel for cooking, warming, and smelting ore; shredded bark for things like twine and diapers; food; medicines; potential rangeland. In fact, whether we're talking about junipers or any other ecosystem, we easily recognize provisioning services, because "provisions" are so obvious in our lives.

We're usually less conscious of ecosystems' regulating and supporting services like carbon storage, oxygen production, habitats, nutrient cycling, shade, and microclimate modification. "Provisions" are king. In fact, we often get our provisions in a way that harms ecosystems and their ability to provide the other services. Like the miners and prehistoric peoples who decimated the very woodlands that they needed, we consume in the short term.

In 2010 Michael Shermer wrote that "life has never been so good for our species." Sure, he said, we've got our problems, but think about how shockingly deprived people used to be. If you went into a hunter-gatherer village and counted up the number of items the *whole village* owned, you would only be able to find three hundred or so things to count. And the average annual "income" of that society would equal about $100. But today! "Villagers" in Manhattan collectively own and use about ten billion material items, and they pull in an average annual income of $40,000. We have more SUVs, DVDs, PCs, TVs, designer clothes, appliances, and gadgets; we have bigger homes, more leisure, and more people with health insurance; and L.A. has less smog—which makes training for bicycle races so much more pleasant than it used to be.[3] Therefore life has never been so good. (If you are one of the billions who don't have much stuff, or health insurance, too bad.)

I looked to see signs of irony in this article but could find none. I myself like stuff, but this view of ever-multiplying "provisions" as the key to the good life seems rather narrow.

*

Speaking of provisioning services, Lincoln County, Nevada, has been working with a Chinese company on a project to cut pinyons and junipers to use for biomass energy production. The idea inspires as much passion among some people as does the practice of exporting shiploads of redwood logs to Japan or

Utah juniper at Dead Horse Point.

China. Supporters point out the economic benefits of the biomass energy, and they also predict better habitat for sage grouse and other wildlife, watershed health, wildfire hazard reduction, increased biodiversity, and more pine nut production. Critics argue that using old-growth forests for energy production is an unsustainable practice. One activist claims that this "expensive, wasteful, wildlife habitat-destroying boondoggle" would "turn large tracts of the Great Basin into a desertified wasteland."[4]

These adversaries are arguing about the value and optimization of various ecosystem services, and what makes for a resilient ecosystem. Provisioning services—energy and the revenues from it—are clearly driving the situation. In this debate, and in general, cultural services carry less weight. The effects of spiritual renewal, recreation, education, and aesthetics are not concrete or quantifiable; they are touchy-feely. No one would argue that these things don't enrich our lives. (And fortunately, when we want to take advantage of cultural services, we have the provisions to facilitate that: backpacks, cars, ATVs, visitor centers, mountain bikes, hiking shoes, camp hammocks, and books about junipers.) However, in today's mainstream value system, cultural services are not on the same level as profits, let alone clean water or oxygen. We look at cultural services as simply nice to have.

As we have seen, though, traditional peoples relied on junipers' cultural gifts as essential. The plant itself provided materials for cleansing and purifying practices in times of birth, sickness, and death. The smoke provided a means of blessing, and people used the plants and animals within the pinyon-juniper woodland for fundamental rituals. The people who used the junipers in these ways were not so different from us. They used—and sometimes abused—provisioning, supporting, and regulating services, just as we do. When population numbers outstripped available resources, their spiritual connection to juniper could not keep them from throwing the ecosystem out of balance as they depleted those resources.

But their spiritual connections to junipers and nature made a difference to their lives. These connections likely restored discouraged spirits, contributed to a sense of balance and wholeness, and nourished community unity. In the midst of our technological culture, many people recognize and seek that balance in their own way. John Muir, the apostle of nature's cultural services, inspired and encouraged people to turn to the natural world. He wrote, "Thousands of tired, nerve-shaken, over-civilized people are beginning to find out that going to the mountains is going home; that wildness is a necessity; and that mountain parks and reservations are useful not only as fountains of timber and irrigating rivers, but as fountains of life. Awakening from . . . the deadly apathy of luxury, they are trying as best they can to mix and enrich their own little ongoings with those of Nature, and to get rid of rust and disease."[5]

In Japan—that harried and crowded country, where big cities employ "shovers" wearing white gloves to pack people into jammed subway cars—the government is wholly on board with this idea. For some years, the Japanese Forestry Agency has been developing special Forest Therapy trails so people can practice *shinrin-yoku*, or "forest bathing." The idea is to get out of the city and into the trees, to take slow time to feel, hear, touch, taste, and look—to truly be present with the inhabitants of the forest. A lot of Japanese (and other) research indicates that shinrin-yoku lowers blood pressure, heart rate, and stress hormones; increases vigor, cognition, and empathy; and reduces anger, anxiety, and depression. Even just smelling the aromatic oils of evergreen trees can significantly strengthen the immune system. Some practitioners report that forest bathing increases their intuition and energy, deepens their relationships, and expands their general sense of happiness.[6] As a friend explained it nonscientifically, lying beneath a tree and looking up fills her with wonder, gratitude, and an unexplainable sense of kinship.

Reality is *interaction*, according to some. Physicist Carlo Rovelli suggests that living and nonliving entities are essentially "nodes of interactions." After all, what is the essence of anything apart from its connections, its encounters with others? He puts it this way: "I am not a thing; I'm a net of interactions with the world around me, with the people who know me, who love me."[7] If so, then I am not just a user of ecosystem services. The services are part of the web of interactions that is me, and I am interrelated with all I have encountered.

When we enter a juniper ecosystem—or seashore, desert, mountain, or grassland—we have the opportunity to perceive this interweaving "body to body." The term "cultural services" doesn't adequately describe this kind of encounter, as we leave behind the technology/information web and the insulated containers where we spend most of our time. Among the junipers, we can consciously reconnect with a more fundamental network. We can renew our fellowship with other organisms and entities—trees, insects, soil, sky, grasses, animals, fungi, rocks, birds, water—and in doing that we may experience them as kin to us at a deep level.

*

Recently, I went up the foothills to sit among scattered old junipers high above the Salt Lake Valley. The scrub oak had dropped most of their leaves and a breeze shivered the dried grasses and the late-blooming asters. The noise of traffic rose up from below, but here the trees stood quiet. Dried purple-brown cones and twigs covered the ground. Citizens down below were going places and doing things. As I sat among the junipers, I let go of all the juniper knowledge, letting curiosity and then the present moment take over. The junipers did what they always do: standing all around, they let me be.

I've come to accept that we—individually and collectively—must deal with facts, controversies, decisions, crises, sorrows, injustice, and drudgery. But in order to deal with them well, we would do well to "go into the forest." For there we experience life's mysterious and interwoven wholeness. We can regain a clarity that is too easily clouded over by today's unending demands and information, books, studies, media, meetings, and posts. Whatever our roles may be in relationship to junipers or any other inhabitants of the world, truly being in the natural world may help us see what we could not see before.

As for me, when I descended the foothills that day, I was more aware, more sane—maybe even a little more kind for a couple of hours—for having sat quietly among my kin.

Notes

Prologue: Strands

1. From an article in the *Deseret News* (Salt Lake City, UT) September 25, 1861, reporting on an exploring party into the Uinta Basin. See Jedediah Smart Rogers, "One 'Vast Contiguity of Waste': Documents from an Early Attempt to Expand the Mormon Kingdom into the Uinta Basin, 1861," *Utah Historical Quarterly* 73 (Summer 2005): 250.
2. John Muir, *My First Summer in the Sierra* (Boston: Houghton Mifflin, 1911), 211.
3. John Tallmadge, "Crazy about Nature," in *The Way of Natural History*, ed. Thomas Lowe Fleischner (San Antonio: Trinity University Press, 2011), 20.
4. Wendell Berry, "The Work of Local Culture," in *What Are People For?* (Berkeley, CA: Counterpoint, 2010), 153–69.
5. Ronald M. Lanner, *The Piñon Pine: A Natural and Cultural History* (Reno: University of Nevada Press, 1981), xii.

Chapter 1: Roots

1. Taxonomic divisions and names may change over time. *Juniperus osteosperma* used to be known as *Juniperus utahensis*, for instance. Some authors state the current scientific name of this tree as *Sabina osteosperma*; for these authors, *Sabina*, which traditionally referred to a group of junipers within the genus *Juniperus*, deserves generic status. *Juniperus* still seems to be the most commonly used scientific name for this genus, however. See "Juniperus," Gymnosperm Database, accessed May 1, 2017, http://www.conifers.org/cu/Juniperus.php.
2. Paul T. Tueller and James E. Clark, "Autecology of Pinyon-Juniper Species of the Great Basin and Colorado Plateau," in *The Pinyon-Juniper Ecosystem: A Symposium*, May 1975 (Logan: Utah State University, College of Natural Resources, Utah Agricultural Experiment Station, 1975), 34–35; Thomas N. Johnsen Jr., "One-Seed Juniper Invasion of Northern Arizona Grasslands," *Ecological Monographs* 32 (Summer 1962): 199.
3. Dorothy Parker, "Interior," in *The Complete Poems of Dorothy Parker* (New York: Penguin, 2003).

4. For junipers as one of two highest-known forests in the world, see Arndt Hampe and Rémy J. Petit, "Cryptic Forest Refugia on the 'Roof of the World,'" *New Phytologist* 185 (December 2009): 5–7.
5. For information on this genus, see "Juniperus," Gymnosperm Database, accessed August 24, 2013, http://www.conifers.org/cu/Juniperus.php.
6. Intermountain Society of American Foresters, "Management of Pinyon-Juniper 'Woodland' Ecosystems: A Position of the Intermountain Society of American Foresters," February 15, 2013, 2, accessed February 24, 2014, http://www.usu.edu/saf/PJWoodlandsPositionStatement.pdf.
7. Donald Culross Peattie, *A Natural History of Western Trees* (Boston: Houghton Mifflin, 1953), 263–64. The name "Deseret" is a Mormon term; when the Utah settlers first petitioned Congress for statehood, they proposed that the name be Deseret. A more well-known comparison between Mormons and trees comes from Wallace Stegner, who suggested that rows of Lombardy poplars in Mormon country "appealed obscurely to the rigid sense of order of the settlers, and that a marching row of plumed poplars was symbolic, somehow, of the planter's walking with God and his solidarity with his neighbors." See Wallace Stegner, *Mormon Country* (New York: Bonanza Books, 1942), 23–24.
8. Ron Halvorsen, "Western Juniper (*Juniperus occidentalis*)," *Kalmiopsis* 20 (2013): 26.
9. Joe Patoski, "The War on Cedar," *Texas Monthly*, December 1997.
10. Jose Salinas and Chip Cartwright, "Welcoming and Opening Remarks," in *Desired Future Conditions for Piñon-Juniper Ecosystems Symposium*. General Technical Report RM-258, tech. coordinators Douglas W. Shaw, Earl F. Aldon, and Carol LoSapio (Fort Collins, CO: U.S. Department of Agriculture, Forest Service, Rocky Mountain Forest and Range Experiment Station, 1995), 3.
11. Edward Abbey, *A Voice Crying in the Wilderness (Vox Clamantis in Deserto): Notes from a Secret Journal* (New York: St. Martin's, 1989).
12. Telephone interview with Ken Yamane, July 10, 2010; telephone interview with Dallas Roberts, November 4, 2014.
13. Native American Ethnobotany database, University of Michigan–Dearborn, http://naeb.brit.org/; Utah state legislature website, http://le.utah.gov; *Iron County Record* (Cedar City, UT), March 23, 1929; *Salt Lake Telegram*, February 25, 1919; February 16, 1923; December 30, 1933; June 11, 1945; *Murray (UT) Eagle*, February 23, 1933.
14. "Jump for the Juniper," accessed September 23, 2010, http://jumpforthejuniper.blogspot.com/.
15. "Minutes of the House Natural Resources, Agriculture, and Environment Standing Committee," February 20, 2008, accessed November 26, 2014, http://le.utah.gov/~2008/minutes/hnae0220.pdf.

16. "Utah State Tree—Quaking Aspen," Utah's Online Library, Utah State Library Division, accessed September 17, 2014, http://pioneer.utah.gov/research/utah_symbols/tree.html.
17. See Jennifer DeWoody, Carol A. Rowe, Valerie D. Hipkins, and Karen E. Mock, "'Pando' Lives: Molecular Genetic Evidence of a Giant Aspen Clone in Central Utah," *Western North American Naturalist* 68 (2008): 493–97; John Hollenhorst, "Central Utah's Pando, World's Largest Living Thing, Is Threatened, Scientists Say," *Deseret News* (Salt Lake City), October 7, 2010.

Chapter 2: Germination

1. Robert P. Adams, *Junipers of the World: The Genus* Juniperus, 4th ed. (Bloomington, IN: Trafford, 2014), 342.
2. Tueller and Clark, "Autecology of Pinyon-Juniper Species," 31.
3. Henry S. Graves, Frank J. Phillips, and Walter Mulford, *Utah Juniper in Central Arizona*, U.S. Department of Agriculture, Forest Service, Circular 197 (June 8, 1912), 5. For years, I could never find a definitive answer on how the holes get into the seeds. Finally, I bought on eBay a strand of beads collected on the Navajo Reservation, and I asked the seller whether rodents or insects made the holes. He wrote back, "It is both. My aunt picks them. She says it's mainly ants that get into them, but sometimes squirrels and other rodents if they are around. Not too many rodents in the area where she gets them. If rodents get to them sometimes they leave disease behind. Too many elders were coming down with that hantavirus that has been going around. So my aunt likes the area she gets them.... The ants like the center of the Juniper Berries because there is something sweet inside they like to get to. I sell for my aunt online. She is not too tech savvy." "Lawrence," e-mail message to author, February 15, 2017.
4. Barre Toelken, "Seeing with a Native Eye: How Many Sheep Will It Hold?," in *Seeing with a Native Eye: Essays on Native American Religion*, ed. Walter Holden Capps (San Francisco: Harper, 1976), 18–19.
5. "Love Beads Cut Welfare Rolls among Utah Indians," *Observer-Reporter* (Washington, PA), September 9, 1968; Richard Movitz obituary, *Salt Lake Tribune*, May 18, 2010.
6. Tueller and Clark, "Autecology of Pinyon-Juniper Species," 31.
7. Neil C. Frischknecht, "Native Faunal Relationships within the Pinyon-Juniper Ecosystem," in *Pinyon-Juniper Ecosystem*, 62.
8. Thomas N. Johnsen Jr., "Longevity of Stored Juniper Seeds," *Ecology* 40 (1959): 487–88, referenced in Elena Zlatnik, *Juniperus osteosperma*, Fire Effects Information System (U.S. Department of Agriculture, Forest Service, Rocky Mountain Research Station, Fire Sciences Laboratory, 1999); accessed June 5, 2010, http://www.fs.fed.us/database/feis/plants/tree/junost/all.html.

9. Edward Abbey, *Desert Solitaire* (New York, Ballantine Books, 1971), 32. Arches National Monument became Arches National Park in 1971.
10. Uzi Avner of the Aravah Institute told colleagues that "in the late Neolithic burial ground of Eilat, I excavated a nicely built installation, with the remains of a Juniper tree inside, a tree-trunk 30 cm high and 14 cm wide. It was dated by ^{14}C [carbon 14] to about 5,000 B.C.E. and clearly represents Asherah remains. There are four other remains from different periods that were suggested to be related to Asherah in the Near East." See William G. Dever and Seymore Gitin, eds., *Symbiosis, Symbolism, and the Power of the Past: Canaan, Ancient Israel, and Their Neighbors from the Late Bronze Age through Roman Palaestina*, Proceedings of the Centennial Symposium, W. F. Albright Institute of Archaeological Research and American Schools of Oriental Research, Jerusalem, May 29–May 31, 2000 (Warsaw, IN: Eisenbrauns, 2003), 35. See also Paul Kendall, "Juniper," Trees for Life, accessed May 6, 2011, http://treesforlife.org.uk/forest/mythology-folklore/juniper/.
11. Susan Ackerman, "At Home with the Goddess," in Dever and Gitin, *Symbiosis, Symbolism*, 456–65. See also Judith M. Hadley, *The Cult of Asherah in Ancient Israel and Judah: Evidence for a Hebrew Goddess*, University of Cambridge Oriental Publications, vol. 57 (Cambridge, 2000), 35; and Paul Kendall, "Juniper," Trees for Life; accessed May 6, 2011, http://treesforlife.org.uk/forest/mythology-folklore/juniper/.
12. Alfred E. P. Raymond Dowling, *The Flora of the Sacred Nativity* (London: Kegan Paul, Trench, Trubner, 1900), 198–99.
13. Ibid., 198.
14. "Ginevra de' Benci," National Gallery of Art, accessed January 17, 2014, http://www.nga.gov/content/ngaweb/Collection/highlights/highlight50724.html.
15. Fred Hagender, *The Meaning of Trees: Botany, History, Healing, Lore* (San Francisco: Chronicle Books, 2005), 119.
16. "Earth Wisdom: Juniper," Puakai Healing, accessed April 17, 2017, http://puakaihealing.com/earth-wisdom-juniper-tree/.
17. Alisa Battaglia, "The Sacred Art of Ritual of Smudging," accessed April 17, 2017, http://www.ecologyofthespirit.com/_infoexchange/articles/Smudging.htm.
18. Kathleen Stokker, *Remedies and Rituals: Folk Medicine in Norway and the New Land* (Saint Paul: Minnesota Historical Society Press, 2007).
19. "Pinkie's Parlour: Juniper," accessed June 8, 2011, http://www.angelfire.com/journal2/flowers/pcd40.html.
20. Patrick C. Morris, "Bears, Juniper Trees, and Deer, the Metaphors of Domestic Life, an Analysis of a Yavapai Variant of the Bear Maiden Story," *Journal of Anthropological Research* 32 (Autumn 1976): 247–48, 252–53.
21. In German, *wach* is a form of the verb meaning "to wake." In earlier usage, however, it also referred to watching over, guarding, and aliveness. *Holder* is

mainly a botanical term, used in some plant names; however, the suffix *-er* is also used to designate someone doing something, as in *Lehrer* (teacher). See Joseph Leonhard Hilpert, *Englisch-Deutsches und Deutsch-Englisches Worterbuch* [A Dictionary of the English and German, and the German and English Language], vol. 2, *L–Z* (Carlsruhe, Germany: Th. Braun, 1845).

22. Hagender, *Meaning of Trees*, 119.
23. Jacob Grimm, *Teutonic Mythology*, vol. 2, trans. James Steven Stallybrass (London: George Bell and Sons, 1883), 872, 884.
24. "Iskanderkul Lake," Tajikistan Travel Guide, accessed January 10, 2014, http://www.traveltajikistan.net/gosee/iskanderkul_lake/.
25. "Reminiscences of John R. Young," *Utah Historical Quarterly* 3 (July 1930): 83.
26. "Street Railroad Extension," *Salt Lake Herald*, April 27, 1876.
27. "Plans Are Nearing Completion for Big Celebration," *Deseret News*, July 7, 1923.
28. "Old Cedar Post Object of Reverence," *Salt Lake Telegram*, August 13, 1924. In 1925, an automobile knocked down the cedar post; the city parks department put it back up. *Salt Lake Telegram*, November 13, 1952.
29. "West Joins Utah in Pioneer Day," *Deseret News*, July 24, 1933.
30. "Vandals Cut Down Historic S.L. Tree," *Deseret News,* September 22, 1958.
31. "Desecration of a Landmark," *Deseret News*, September 25, 1958.
32. A. R. Mortensen, "In Memoriam: Kate Carter," *Utah Historical Quarterly* 44 (Fall 1976): 395–96. For a detailed account of this event, see Gary Topping, "One Hundred Years at the Utah State Historical Society," *Utah Historical Quarterly* 65 (Summer 1997): 265–71.
33. Roger Roper, historic preservation coordinator for the Utah Division of State History, found this article and brought it to my attention some years ago.

Chapter 3: Survival

1. Larry T. DeBlander, John D. Shaw, Chris Witt, Jim Menlove, Michael T. Thompson, Todd A. Morgan, R. Justin DeRose, and Michael C. Amacher, *Utah's Forest Resources, 2000–2005*, Resource Bulletin 777 RMRS-RB-10 (Fort Collins, CO: U.S. Department of Agriculture, Forest Service, Rocky Mountain Research Station, December 2010), 11; Renee A. O'Brien, *Arizona's Forest Resources, 1999*, Resource Bulletin RMRS-RB-2 (Fort Collins, CO: U.S. Department of Agriculture, Forest Service, Rocky Mountain Research Station, January 2002), 35; J. David Born, Ronald P. Tymcio, and Osborne E. Casey, *Nevada's Forest Resources*, Resource Bulletin INT-76 (Fort Collins, CO: U.S. Department of Agriculture, Forest Service, Intermountain Research Station, July 1992), 49; Sara A. Goeking, John D. Shaw, Chris Witt, Michael T. Thompson, Charles E. Werstak Jr., Michael C. Amacher, Mary Stuever, et al., *New Mexico's Forest Resources, 2008–2012*, Resource Bulletin RMRS-RB-18 (Fort Collins, CO: U.S. Department of Agriculture, Forest Service, Rocky Mountain Research Station, August 2014),

103; Neil E. West, Kenneth H. Rea, and Robin J. Tausch, "Basic Synecological Relationships in Juniper-Piñon Woodlands," in *Pinyon-Juniper Ecosystem*, 42. Pygmy-conifer woodlands (often, but not necessarily, juniper) also grow on the Great Plains, Columbia Plateau, and Pacific Border, and in the Rocky Mountains, Wyoming Basin, Sierra Nevada, and Cascade Range.

2. Zlatnik, *Juniperus osteosperma.*
3. Janette S. Scher, *Juniperus scopulorum*, Fire Effects Information System, accessed June 5, 2010.
4. Diary of William H. Ashley, March 25–June 27, 1825, William H. Ashley's 1825 Rocky Mountain Papers, Library of Western Fur Trade Historical Source Documents, accessed January 24, 2014, http://user.xmission.com/~drudy/mtman/html/ashintro.html. Boat information comes from Ann Zwinger, *Run, River, Run: A Naturalist's Journey Down One of the Great Rivers of the American West* (Phoenix: University of Arizona Press, 1975), 78.
5. Cynthia J. Willson, Paul S. Manos, and Robert B. Jackson, "Hydraulic Traits Are Influenced by Phylogenetic History in the Drought-Resistant, Invasive Genus *Juniperus* (Cupressaceae)," *American Journal of Botany* 95 (March 2008): 307.
6. Robert P. Adams, *Junipers of the World*, 81.
7. Janet Burton Seegmiller, *A History of Iron County* (Salt Lake City: Utah State Historical Society and Iron County Commission, 1998), 34.
8. Dale Morgan, *Jedediah Smith and the Opening of the West* (Lincoln, NE: Bison Books, 1964), 211–13.
9. Willson, Manos, and Jackson, "Hydraulic Traits," 229–300, 307; M. J. Linton, J. S. Sperry, and D. G. Williams, "Limits to Water Transport in *Juniperus osteosperma* and *Pinus edulis:* Implications for Drought Tolerance and Regulation of Transpiration," *Functional Ecology* 12, no. 6 (1998): 906–11. The droughts of 1996 and 2002, for instance, demonstrated the susceptibility of pinyons and the resistance of junipers to dry conditions.
10. "Why Juniper Trees Can Live on Less Water," *Duke Today*, February 27, 2008, accessed April 30, 2010, http://today.duke.edu/2008/02/cedardry.html.
11. Willson, Manos, and Jackson, "Hydraulic Traits," 299.
12. "Why Juniper Trees Can Live on Less Water."
13. Abbey, *Desert Solitaire*, 265.
14. Edwin Bryant, *What I Saw in California, Being the Journal of a Tour, in the Years 1846, 1847* (New York: D. Appleton, 1849), 144.
15. Present-day Redlum Spring. See Henry J. Webb, "Edwin Bryant's Trail through Western Utah," *Utah Historical Quarterly* 29 (April 1961): 131–32.
16. Bryant, *What I Saw in California*, 170–72.
17. William B. Smart and Donna T. Smart, *Over the Rim: The Parley P. Pratt Exploring Expedition to Southern Utah, 1849–1850* (Logan: Utah State University Press, 1999), 51, 59, 65, 77, 109.

18. "Memoria in Aeterna," *Deseret News*, June 3, 1893.
19. Karl Young, "Wild Cows of the San Juan," *Utah Historical Quarterly* 32 (July 1964): 254–55.
20. W. L. Rusho, *Everett Ruess: A Vagabond for Beauty and Wilderness Journals* (Layton, UT: Gibbs Smith, 2002), 191–92.
21. John D. Barton, *A History of Duchesne County* (Salt Lake City: Utah State Historical Society and Duchesne County Commission, 1998), 117.

Chapter 4: Spirals

1. T. T. Kozlowski, *Growth and Development of Trees*, vol. 2, *Cambial Growth, Root Growth, and Reproductive Growth* (New York: Academic Press, 1971), 79.
2. Robert W. Chambers, "Spiral Grain: The Inside Story," *Log Building News*, May/June/July 2007, 7–8, accessed July 10, 2011, http://www.logassociation.org/resources/spiral_grain_lbn63.pdf.
3. See Sondre Skatter and Bohumil Kucera, "The Cause of the Prevalent Directions of the Spiral Grain Patterns in Conifers," *Trees* 12, no. 5 (1998): 265–73.
4. Kozlowski, *Growth and Development of Trees*, 80–81; Chambers, "Spiral Grain," 7–8. See also K. Schulgasser and A. Witztum, "The Mechanism of Spiral Grain Formation in Trees," *Wood Science Technology* 41, no. 2 (2007): 133–56.
5. Theodore Andrea Cook, *The Curves of Life* (London: Constable, 1914), viii.
6. Schulgasser and Witztum, "Mechanism of Spiral Grain Formation," 133–56. For another discussion of this topic, see Christopher J. Earle, "Why Do Trees Form Spiral Grain?," Gymnosperm Database, accessed February 17, 2014, http://www.conifers.org/topics/spiral_grain.php.
7. Theophrastus, *Enquiry into Plants and Minor Works on Odours and Weather Signs*, trans. Arthur Hort (Cambridge, MA: Harvard University Press, 1916), 427.
8. For the last two, see Hans Kubler, "Function of Spiral Grain in Trees," *Trees* 5 (1991), 125–35.
9. Chambers, "Spiral Grain," 8.
10. John Burroughs, *The Writings of John Burroughs*, vol. 17 (Boston: Houghton Mifflin, 1913), 86.
11. This idea was expressed by Cook, *Curves of Life*, 413.
12. Andrea Brunelle, "An Introduction to Paleoecological Data and Their Utility in Ecosystem Restoration" (PowerPoint presentation at the conference "Restoring the West 2009—Peaks to Valleys: Innovative Land Management for the Great Basin," Logan, UT, October 28, 2009), accessed April 30, 2010, https://connect.usu.edu/p58734826/.
13. Frischknecht, "Native Faunal Relationships," 59.

14. Larry L. Coats, Kenneth L. Cole, and Jim I. Mead, "50,000 Years of Vegetation and Climate History on the Colorado Plateau, Utah and Arizona, USA," *Quaternary Research* 70, no. 2 (2008): 322–38.
15. Cheryl L. Nowak, Robert S. Nowak, Robin J. Tausch, and Peter E. Wigand, "Tree and Shrub Dynamics in Northwestern Great Basin Woodland and Shrub Steppe during the Late-Pleistocene and Holocene," *American Journal of Botany* 88 (March 1994): 265–77.
16. See "Condor Viewing in the Vermilion Cliffs National Monument," U.S. Department of the Interior, Bureau of Land Management, accessed August 20, 2014, http://www.blm.gov/az/st/en/prog/recreation/watchable/condors.html.
17. Joseph M. Trudeau, *An Environmental History of the Kane and Two Mile Ranches in Arizona* (report prepared for the Grand Canyon Trust, 2006), 22.
18. The juniper mistletoe is *Phoradendron juniperinum.* See Colorado State University Extension website, accessed August 20, 2014, http://www.ext.colostate.edu/pubs/garden/02925.html.
19. Catharine Gehring and Thomas Whitham, "Environmental Stress Influences Aboveground Pest Attack and Mycorrhizal Mutualism in Piñon-Juniper Woodlands: Implications for Management in the Event of Global Warming," in Shaw, Aldon, and LoSapio, *Desired Future Conditions*, 30–37.
20. Julio L. Betancourt, Elizabeth A. Pierson, Kate A. Rylander, James A. Fairchild-Parks, and Jeffrey S. Dean, "Influence of History and Climate on New Mexico Piñon-Juniper Woodlands," in *Managing Piñon-Juniper Ecosystems for Sustainability and Social Needs*, tech. coordinators Earl F. Aldon and Douglas W. Shaw, General Technical Report RM-236 (Fort Collins, CO: U.S. Department of Agriculture, Forest Service, Rocky Mountain Forest and Range Experiment Station, 1993), 46–48.

    According to Betancourt et al., it appears that from 5000 to 3500 BP, precipitation increased again. After that, from 3500 to 2600 BP, a neoglacial period cooled the West, causing the juniper woodlands to move again to lower elevations and expand. But the cycle came around again, and when precipitation fell off during a drought between 2600 and 1000 BP, the juniper woodlands again moved to higher terrain and latitudes. Desert shrubs moved into the places they left behind. Once more, a few hundred years ago, the earth again entered a cooler period. The Little Ice Age, as it is called, lasted until the mid-nineteenth century. Besides making life difficult and precarious for human populations, the colder and wetter conditions of the Little Ice Age again changed landscapes. Vegetation types shifted to lower elevations once again. During all this time, fire did its own work on the landscape.
21. "Dynamics of Utah Juniper Woodlands in Wyoming: Packrat Middens," n.d., accessed May 7, 2010, http://wwwpaztcn.wr.usgs.gov/wyoming/middens.html.

Scientific evidence is not always consistent. Researchers in another study concluded that Utah juniper colonized East Pryor Mountain between 7500 and 5400 years BP, during a rather dry period. See Mark E. Lyford, Stephen T. Jackson, Julio L. Betancourt, and Stephen T. Gray, "Influence of Landscape Structure and Climate Variability on a Late Holocene Plant Migration," *Ecological Monographs* 73, no. 4 (2003): 567.

22. Steve Jackson and Julio Betancourt, *Late Holocene Expansion of Utah Juniper in Wyoming: A Modeling System for Studying Ecology of Natural Invasions*, NSF-DEB-9815500, Final Report, Collaborative Research (National Science Foundation, n.d.), 2. The cooling and warming cycles I give are just a coarse overview, from a limited point of view. Some scientists today believe that the earth experienced sixteen to eighteen glacial cycles during the Pleistocene, the two and a half million years before our own geologic epoch, the Holocene. During these cycles, the colder periods—when glaciers covered the northern latitudes—lasted much longer than the intervening warming periods. Also, it appears that regional climate changes could occur swiftly. See Betancourt et al., "Influence of History and Climate," 42–43.
23. Robert Smithson, *Robert Smithson: The Collected Writings* (Berkeley: University of California Press, 1996), 146–47.
24. Betancourt et al., "Influence of History and Climate," 42–56.
25. See Thomas R. Cartledge and Judith G. Propper, "Piñon-Juniper Ecosystems through Time: Information and Insights from the Past," in Aldon and Shaw, *Managing Piñon-Juniper Ecosystems*, 69–70.

Chapter 5: Leaves and Seeds

1. Jackson Browne began writing this song, basing it on an experience he had in Flagstaff, Arizona. Glenn Frey of the Eagles finished it, changing the venue to Winslow, and the Eagles recorded it in 1972.
2. See Adriana Rissetto, "Between Four Sacred Mountains: The Diné and the Land in Contemporary America" (website developed at the University of Virginia in fulfillment of Master of Arts requirements, 1997), accessed March 21, 2014, http://xroads.virginia.edu/~ma97/dinetah/change2.html.
3. Georgiana Kennedy Simpson, *Navajo Ceremonial Baskets: Sacred Symbols, Sacred Space* (Summertown, TN: Native Voices, 2003), 15–17.
4. For fascinating microscopic close-up photos of juniper leaves, see Coats, Cole, and Mead, "50,000 Years," 327.
5. See Willson, Manos, and Jackson, "Hydraulic Traits," 307.
6. J. T. Liddell, "The Very Spicy Air of Juniper Lures Utahns Afield," *Deseret News*, December 19, 1959. If the sentence beginning "No other tree" sounds familiar,

it's because I have already quoted it from another source; the reporter lifted it from Peattie, *Natural History of Western Trees*, 264.

7. Karen R. Adams, "Subsistence and Plant Use during the Chacoan and Second Occupations at Salmon Ruin," in *Chaco's Northern Prodigies: Salmon, Aztec and the Ascendance of the Middle San Juan Region after AD 1100*, ed. Paul F. Reed (Salt Lake City: University of Utah Press, 2008), 75.
8. Ibid., 79, 81.
9. Nicholas Culpeper, *Culpeper's Complete Herbal: With Nearly Four Hundred Medicines, Made from English Herbs, Physically Applied to the Cure of All Disorders Incident to Man; with Rules for Compounding Them: Also, Directions for Making Syrups, Ointments, &C* (London: Milner and Sowerby, 1852), 206–7.
10. See Ernest Small, *North American Cornucopia: Top 100 Indigenous Food Plants* (Boca Raton, FL: CRC Press, 2013), 417.
11. Native American Ethnobotany Database, accessed March 24, 2014, http://herb.umd.umich.edu/.
12. Dorena Martineau, e-mail message to author, May 23, 2010.
13. Daniel Moerman, *Native American Ethnobotany* (Portland, OR: Timber Press, 1998). The WebMD website confirms the use of juniper for many of these purposes, and more. It also discusses side effects and possible harmful effects; accessed August 21, 2014, http://www.webmd.com/vitamins-supplements/ingredientmono-724-juniper.aspx?activeingredientid=724&activeingredientname=juniper.
14. Toelken, "Seeing with a Native Eye," 15.
15. Native American Ethnobotany Database.
16. Henry F. Dobyns, *Spanish Colonial Tucson: A Demographic History* (Tucson: University of Arizona Press, 1976), 26–27.
17. Elliott Coues, ed. and trans., *On the Trail of a Spanish Pioneer: The Diary and Itinerary of Francisco Garcés*, vol. 2 (New York: Francis P. Harper, 1900), 329, 344–47. Note that A. W. Whipple, traveling through Arizona in 1854, describes large cedar trees bearing sweet berries. Harley G. Shaw, *Wood Plenty, Grass Good, Water None: Vegetation Changes in Arizona's Upper Verde River Watershed from 1850 to 1997*, General Technical Report RMRS-GTR-177 (U.S. Department of Agriculture, Forest Service, Rocky Mountain Research Center, 2006), 12–16.
18. See "Nutritional Mineral Supplements from Plant Ash," U.S. patent application 20040126460 A1, accessed August 21, 2014, http://www.google.com/patents/US20040126460. See also Vernon O. Mayes and Barbara Bayless Lacy, *Nanise': A Navajo Herbal; One Hundred Plants from the Navajo* (Tsaile, AZ: Navajo Community College Press, 1989).
19. Shirlee Silversmith, personal communication, January 24, 2013.

20. See recipes provided by the Valley Mills company, accessed August 21, 2014, http://valleymillsbluecorn.com/Recipes.html.
21. Cartledge and Propper, "Piñon-Juniper Ecosystems," 68.
22. Zlatnik, *Juniperus osteosperma*.
23. Frischknecht, "Native Faunal Relationships," 106.
24. Ibid., 116.
25. Ibid., 113. Also Ted L. Terrel and J. Juan Spillet, "Pinyon-Juniper Conversion: Its Impact on Mule Deer and Other Wildlife," in *Pinyon-Juniper Ecosystem*, 105–17.
26. "Rabbits Threaten Our Gin and Tonic: Wild Animals Are Overgrazing on Juniper Berries," *Daily Mail*, accessed March 19, 2014, http://www.dailymail.co.uk/news/article-2543711/Rabbits-threatens-gin-tonic-Wild-animals-overgrazing-juniper-berries.html.
27. Lena K. Ward and Catharine H. Shellswell, *Looking after Juniper: Ecology, Conservation and Folklore* (Salisbury, England: Plantlife, 2017), 5, accessed April 21, 2017, http://www.plantlife.org.uk/application/files/7614/8958/6210/JUNIPER_DOSSIER_13_2_17_CS.pdf.
28. Terrel and Spillet, "Pinyon-Juniper Conversion," 117.

Chapter 6: Wood

1. Leonard E. Read, "I, Pencil: My Family Tree as Told to Leonard E. Read" (first published in *The Freeman*, December 1958), Library of Economics and Liberty, accessed September 3, 2014, http://www.econlib.org/library/Essays/rdPncl1.html. This essay is a classic of conservative economic thought valorizing Adam Smith's "Invisible Hand," without considering attendant environmental and social issues. However, beyond its political views, Read's remark about a tree is undeniably true.
2. Tueller and Clark, "Autecology of Pinyon-Juniper Species," 29.
3. Zlatnik, *Juniperus osteosperma*.
4. Ronald Lanner, *Trees of the Great Basin: A Natural History* (Reno: University of Nevada Press, 1984), 112–13.
5. Quoted in William Rossi, ed., *Wild Apples and Other Natural History Essays by Henry D. Thoreau* (Athens: University of Georgia Press, 2002), xii.
6. Roland Emos, "Trees: Magnificent Structures," U.K. Natural History Museum web page, accessed March 4, 2014, http://www.nhm.ac.uk/print-version/?p=/nature-online/life/plants-fungi/magnificent-trees/session3/index.html.
7. See *Online Etymology Dictionary*, accessed March 25, 2014, http://www.etymonline.com/index.php; also Dictionary.com, http://dictionary.reference.com, and WordReference.com, http://www.wordreference.com/.
8. Karen R. Adams, "Subsistence and Plant Use," 79.

9. Ibid.
10. See Thomas C. Windes and Eileen Bacha, "Sighting along the Grain: Differential Wood Use at Salmon Ruin," in Reed, *Chaco's Northern Prodigies*, 113–39.
11. Karen R. Adams, "Subsistence and Plant Use," 77–78, 82.
12. Ibid., 80.
13. Peattie, *Natural History of Western Trees*, 265. Noel Morss found "beds of cedar bark" in rock structures during his 1920s investigation of the Fremont culture in the Capitol Reef area of Utah. See Miriam B. Murphy, *A History of Wayne County* (Salt Lake City: Utah State Historical Society and Wayne County Commission, 1999), 34.
14. Joel Janetski, Karen D. Lupo, John M. McCullough, and Shannon A. Novak, "The Mosida Site: A Middle Archaic Burial from the Eastern Great Basin," *Journal of California and Great Basin Anthropology* 14, no. 2 (1992): 183–86.
15. Robert P. Adams, "Yields and Seasonal Variation of Phytochemicals from *Juniperus* Species of the United States," *Biomass* 12, no. 2 (1987): 129–39.
16. Enos A. Mills, *The Story of a Thousand-Year Pine* (Boston: Houghton Mifflin, 1909), 6; Robert P. Adams, "Yields and Seasonal Variation," 129–39.
17. Zlatnik, *Juniperus osteosperma*; Tueller and Clark, "Autecology of Pinyon-Juniper Species," 34; Peattie, *Natural History of Western Trees*, 265.
18. James A. Erdman, "Pinyon-Juniper Succession after Natural Fires on Residual Soils of Mesa Verde, Colorado," Pamphlet 13112, Utah History Research Center (Provo, UT: Brigham Young University, 1970), 4, 14–16. See also Christopher J. Earle, "How Old Is That Tree?," Gymnosperm Database, accessed March 6, 2014, http://www.conifers.org/topics/oldest.htm.
19. Peter M. Brown, OLDLIST database, Rocky Mountain Tree-Ring Research, accessed July 13, 2015, http://www.rmtrr.org/oldlist.htm. This same database indicates that a few years ago, a living bristlecone pine in the White Mountains was found to be the oldest known tree, aged 5,062 years in 2012—a new record.
20. Native American Ethnobotany Database.
21. Dorena Martineau, e-mail message to author, May 23, 2010.
22. Mayes and Lacy, *Nanise'*.
23. Howard Stansbury, *An Expedition to the Valley of the Great Salt Lake of Utah* (Philadelphia: Lippincott, Grambo, 1855), Making of America digital library, University of Michigan, accessed September 9, 2014, http://quod.lib.umich.edu/cgi/t/text/text-idx?c=moa;idno=AJA3655.
24. A. R. Mortensen, ed., "Journal of John A. Widtsoe, Colorado River Party, September 3–19, 1922," *Utah Historical Quarterly* 23 (1955): 228. Widtsoe's journey was mainly for the purpose of attending meetings in Santa Fe to help hammer out the Colorado River Compact, a document allocating water from the

Colorado River among several western states. The compact has had a huge impact on western states (as well as the river!), but that is another story.

25. Linda King Newell and Vivian Linford Talbot, *A History of Garfield County* (Salt Lake City: Utah State Historical Society and Garfield County Commission, 1998), 74.
26. See Angus M. Woodbury, "A History of Southern Utah and Its National Parks," *Utah Historical Quarterly* 12 (July/October 1944): 157.
27. Edward A. Geary, *A History of Emery County* (Salt Lake City: Utah State Historical Society and Emery County Commission, 1996), 181.
28. Martha Sonntag Bradley, *A History of Beaver County* (Salt Lake City: Utah State Historical Society and Beaver County Commission, 1999), 196–97; Jay M. Haymond, interview with Bill Wood, Charles K. Jamison, and Randall M. Banks, February 24, 1977.
29. Karl Young, "Wild Cows of the San Juan," *Utah Historical Quarterly* 32 (July 1964): 254.
30. Levi Peterson, "Juanita Brooks: My Subject, My Sister," *Dialogue* 22 (Spring 1989): 18.
31. Brant and Betty Wall, interview with author, May 21, 2010.
32. Lillian Barrus Nelson, *Juniper and Black Pine: A History of Two Southern Idaho Communities, 1870s to 1995* (Salt Lake City: Publishers Press, 1996), 39.
33. Ibid., 203.
34. Joseph M. Trudeau, "An Environmental History of the Kane and Two Mile Ranches in Arizona" (report prepared for the Grand Canyon Trust, 2006), 72, accessed May 6, 2014, http://www.grandcanyontrust.org/sites/default/files/resources/Environmental_History_Kane_and_Two_Mile_Ranches.pdf.
35. Carl M. Johnson, "Pinyon-Juniper Forests: Assets or Liability," in *Pinyon-Juniper Ecosystem*, 121.
36. Patoski, "War on Cedar."
37. Ibid.
38. Charlie Arvid, quoted in Nancy Turner, *The Earth's Blanket: Traditional Teachings for Sustainable Living* (Seattle: University of Washington Press, 2005), 74–75. Neoanimists put it this way: "The world is full of persons, only some of whom are human, and . . . life is always lived in relationship to others." See Graham Harvey, *Animism: Respecting the Living World* (New York: Columbia University Press, 2005), xi. William Wordsworth expressed a sense of this aliveness in "Ode: Intimations of Immortality from Recollections of Early Childhood," in James Baldwin, *Six Centuries of English Poetry*, Kindle edition (Boston: Silver, Burdett, 1892).
39. Albert Schweitzer, "The Ethics of Reverence for Life," *Christendom* 1 (1936): 225–36.

40. Wordsworth, "Ode: Intimations of Immortality."

Chapter 7: Smoke

1. Helene Gill, ed., *A Creative Approach to Environmental Education: A Teaching Resource Kit for Mountain Countries, Teacher's Manual* (Paris: UNESCO, 2010), 101.
2. Abbey, *Desert Solitaire*, 14.
3. Francis H. Elmore, *Ethnobotany of the Navajo*, Monograph No. 8 (Santa Fe: University of New Mexico and School of American Research, July 1944), 28.
4. William D. Hurst, "Management Strategies within the Pinyon-Juniper Ecosystem," in *Pinyon-Juniper Ecosystem*, 188.
5. Wall, interview.
6. Fred Esplin, interview with author, September 30, 2014; William B. Smart, interview with author, October 5, 2014.
7. Escalante's journal can be found in the Early Americas Digital Archive, published by the Maryland Institute for Technology in the Humanities, University of Maryland, accessed May 29, 2014, http://mith.umd.edu//eada/gateway/diario/diary.html.
8. Gerald W. Williams, comp., "References on the American Indian Use of Fire in Ecosystems" (Washington, D.C.: U.S. Department of Agriculture, Forest Service, 2005).
9. Quoted in David J. Strohmaier, *Drift Smoke: Loss and Renewal in a Land of Fire* (Reno: University of Nevada Press, 2005), 35–36. See also U.S. Geographical and Geological Survey of the Rocky Mountains, "Map of Utah Territory Representing the Extent of the Irrigable, Timber and Pasture Lands, 1878," Cline Library, Special Collections and Archives, Northern Arizona University, http://archive.library.nau.edu/u?/cpa,61301. This map published by the USGS shows large areas where Powell's surveyors either surmised or observed that fire had destroyed trees.
10. William L. Baker and Douglas J. Shinneman, "Fire and Restoration of Piñon-Juniper Woodlands in the Western United States: A Review," *Forest Ecology and Management* 189 (2004): 1–21.
11. William H. Romme, Craig D. Allen, John D. Bailey, William L. Baker, Brandon T. Bestelmeyer, Peter M. Brown, Karen S. Eisenhart, et al., "Historical and Modern Disturbance Regimes, Stand Structures, and Landscape Dynamics in Piñon-Juniper Vegetation of the Western United States," *Rangeland Ecology and Management* 62 (May 2009): 207–9; Baker and Shinneman, "Fire and Restoration," 17. In reviewing the literature on fire and PJ, these authors point to one case where fires prevented tree invasion into sagebrush and grasslands

prior to settlement. In another case, however, the absence of fire did not lead to invasion.

12. Baker and Shinneman, "Fire and Restoration," 9–14, 17. See also John M. Bauer and Peter J. Weisberg, "Fire History of a Central Nevada Pinyon-Juniper Woodland," *Canadian Journal of Forest Research* 39 (August 2009): 1589–99; Romme et al., "Historical and Modern Disturbance," 207–9.
13. Paul A. Arendt and William L. Baker, "Northern Colorado Plateau Piñon-Juniper Woodland Decline over the Past Century," *Ecosphere* 4 (August 2013): 103.
14. See S. F. Arno, "Ecological Effects and Management Implications of Indian Fires," in *Proceedings: Symposium and Workshop on Wilderness Fire, Missoula, Montana, November 15–18, 1983*, ed. J. E. Lotan, B. M. Kilgore, W. C. Fischer, and R. W. Mutch, 82–83, General Technical Report INT-182 (U.S. Department of Agriculture, Forest Service, Intermountain Forest and Range Experiment Station, 1985).
15. D. A. Hester, "The Piñon-Juniper Fuel Type Can Really Burn," Fire Control Notes 13 (U.S. Department of Agriculture, Forest Service, 1951), 26–29, quoted in Baker and Shinneman, "Fire and Restoration."
16. See "Living with Fire in the Pinyon-Juniper Woodland" (pamphlet sponsored by federal, state, and private agencies), accessed September 5, 2014, http://www.unce.unr.edu/publications/files/nr/2003/cm0301.pdf.
17. Alicia L. Reiner, "Fuel Load and Understory Community Changes Associated with Varying Elevation and Pinyon-Juniper Dominance" (master's thesis, University of Nevada–Reno, 2004), 53.
18. Erdman, "Pinyon-Juniper Succession."
19. Morris A. Shirts, "The Iron Mission," in *Utah History Encyclopedia*, ed. Allan Kent Powell (Salt Lake City: University of Utah Press, 1994), 275–76. See also Richard H. Jackson, "Utah's Harsh Lands, Hearth of Greatness," *Utah Historical Quarterly* 49 (Winter 1981): 12–15.
20. For the Iron City story, see Kerry William Bate, "Iron City, Mormon Mining Town," *Utah Historical Quarterly* 50 (Winter 1983): 47–58.
21. Kathy Walker, "Charcoal Ovens Headed for Destruction or Preservation?," *Millard County Chronicle* (Fillmore, UT), March 27, 2003.
22. Scotty Strachan, Franco Biondi, Susan G. Lindström, Robert McQueen, and Peter E. Wigand, "Application of Dendrochronology to Historical Charcoal-Production Sites in the Great Basin, United States," *Historical Archaeology* 47, no. 4 (2013): 103–5.
23. Nell Murbarger, "Charcoal: The West's Forgotten Industry," *Desert*, June 1956, 5–6. A bushel of charcoal might fetch a price as low as eight cents, the amount the Oregon Iron Works paid in 1867; this adds up to a measly two dollars for

every cord of wood cut, hauled, burned, and delivered. At the same time, the smelters at Oreana, Nevada, where trees were more scarce, were paying sixty-five cents per bushel.

24. Thomas J. Straka and Robert H. Wynn, "History on the Road: Charcoal and Nevada's Early Mining History," *Forest History Today*, Fall 2008, 63–65.
25. Strachan et al., "Application of Dendrochronology," 105.
26. Straka and Wynn, "History on the Road," 63–65.
27. Ronald M. Lanner and Penny Frazier, "Historical Stability of Nevada's Pinyon-Juniper Forest," *Phytologia* 93 (December 2011): 373–75.
28. Dongwook W. Ko, Ashley D. Sparrow, and Peter J. Weisberg, "Land-Use Legacy of Historical Tree Harvesting for Charcoal Production in a Semi-arid Woodland," *Forest Ecology and Management* 261 (April 2011): 1283–92.
29. Lanner and Frazier, "Historical Stability," 373.
30. Barbara Tellman and Diana Hadley, *Crossing Boundaries: An Environmental History of the Upper San Pedro River Watershed, Arizona and Sonora* (Tucson: Office of Ethnohistorical Research, Arizona State Museum, University of Arizona, 2006), 19.
31. Mayes and Lacy, *Nanise'*, 55.
32. Small, *North American Cornucopia*, 418–19.
33. Dorena Martineau, e-mail message to author, May 23, 2010.
34. Mark M. Jarzombek, *Architecture of First Societies: A Global Perspective* (Hoboken, NJ: John Wiley & Sons, 2013), 68–69.
35. See Jeremy A. Black, Graham Cunningham, Eleanor Robson, and Gábor Zólyomi, trans., *The Literature of Ancient Sumer* (Oxford: Oxford University Press, 2004), 49. In a manuscript the translator calls "The Building of Ningirsu's Temple," a Sumerian ruler (Gudea) communicates with the gods thus: "He put juniper, the mountains' pure plant, onto the fire, and raised smoke with cedar resin, the scent of gods."
36. Kendall, "Juniper"; Madonna Gauding, *The Signs and Symbols Bible* (New York: Sterling, 2009), 286.
37. David Stern, "Masters of Ecstasy," *National Geographic* 222 (December 2012): 116–18.
38. Moerman, *Native American Ethnobotany*; Elmore, *Ethnobotany of the Navajo*, 28–29, 34, 39.
39. Elmore, *Ethnobotany of the Navajo*, 7–19.
40. "Traditional Navajo Taboos," NavajoCentral.org, accessed May 29, 2014, http://www.navajocentral.org/navajotaboos/taboos_ghosts.html.
41. Moerman, *Native American Ethnobotany*; Elmore, *Ethnobotany of the Navajo.*
42. Peattie, *Natural History of Western Trees*, 277.

43. Jessa Fisher and Phyllis Hogan, "Juniper, the Southwest Tree of Life," Winter Sun, accessed July 20, 2014, http://www.wintersun.com/site/page?view=_article&id=3.
44. "Juniper," *Oxford English Dictionary* (Oxford: Oxford University Press, updated March 2017), accessed at Salt Lake Public Library website, April 24, 2017, http://www.oed.com.ezproxy.slcpl.org/view/Entry/102086?redirectedFrom=JUNIPER#eid.

CHAPTER 8: SPREAD

1. Cartledge and Propper, "Piñon-Juniper Ecosystems," 67–69.
2. See David E. Miller, *Hole in the Rock: An Epic Colonization of the Great American West* (Salt Lake City: University of Utah Press, 1966), 86; William B. Smart, *Old Utah Trails* (Salt Lake City, Utah Geographic Series, 1988), 122–34; and Graig Taylor, "George Brigham Hobbs," Hole in the Rock Foundation, accessed December 5, 2014, http://www.hirf.org/history_bio_hobbs.asp.
3. Harley G. Shaw, *Wood Plenty, Grass Good*, 29.
4. Ibid., 30.
5. Ibid., 19.
6. Ibid., 15–16.
7. Rick Miller, Jeffrey Rose, Tony Svejcar, Jon Bates, and Kara Paintner, "Western Juniper Woodlands: 100 Years of Plant Succession," in Douglas W. Shaw, Aldon, and LoSapio, *Desired Future Conditions*, 50.
8. Bradford P. Wilcox and Thomas L. Thurow, Preface to "Emerging Issues in Rangeland Ecohydrology," special issue, *Hydrological Processes* 20, no. 15 (2006): 3156.
9. T. L. Deboodt, M. P. Fisher, J. C. Buckhouse, and J. Swanson, "Monitoring Hydrological Changes Related to Western Juniper Removal: A Paired Watershed Approach," in *Planning for an Uncertain Future—Monitoring, Integration, and Adaptation*, ed. Richard M. T. Webb and Darius J. Semmens, 227–32. Proceedings of the Third Interagency Conference on Research in the Watersheds, September 8–11, 2008, Estes Park, CO. Scientific Investigations Report 2009-5049 (Reston, VA: U.S. Geological Survey, 2009).
10. Old Growth Western Juniper website, accessed July 2, 2014, http://www.oldgrowthjuniper.com/index.html.
11. Ken Brunges, interview by Ed Brick, October 15, 2010, accessed July 2, 2014, https://www.youtube.com/watch?v=UGzJ4Rw8kXs.
12. Esplin, interview.
13. Romme et al., "Historical and Modern Disturbance," 210–11.
14. Ibid., 203, 213; William H. Romme, Lisa Floyd-Hanna, and David D. Hanna, "Ancient Piñon-Juniper Forests of Mesa Verde and the West: A Cautionary

Note for Forest Restoration Programs," in *Fire, Fuel Treatments, and Ecological Restoration: Conference Proceedings, April 16–18, 2002*, ed. Philip N. Omi and Linda A. Joyce, 345–48, Proceedings RMRS-P-29 (Fort Collins, CO: U.S. Department of Agriculture, Forest Service, Rocky Mountain Research Station, 2003).

15. Malchus B. Baker Jr., Leonard F. DeBano, and Peter F. Folliott, "Soil Loss in Piñon-Juniper Ecosystems and Its Influence on Site Productivity and Desired Future Condition," in Douglas W. Shaw, Aldon, and LoSapio, *Desired Future Conditions*, 10.
16. Darren McAvoy, "USU Researcher Documents Utah's Forest History with Repeat Photographs," *Utah Forest News* 13 (Summer 2009); "Ground Photography: Historical Photography of Twentieth Century Vegetation Change in Wyoming and Montana," USGS National Research Program, accessed November 30, 2016, http://wwwpaztcn.wr.usgs.gov/wyoming/rpt_ground.html; Harley G. Shaw, *Wood Plenty, Grass Good*, 34–42; Stephen A. Hall, William Penner, and Moira Ellis, "Arroyo Cutting and Vegetation Change in Abo Canyon, New Mexico: Evidence from Repeat Photography along the Santa Fe Railway," in *New Mexico Geological Society Guidebook*, 60th Field Conference (2009), 429–38; Utah's Rangelands photo database, Utah State University Extension, accessed October 14, 2014, http://extension.usu.edu/rra/index.htm. See also Romme et al., "Historical and Modern Disturbance," 213.
17. William Hurst, quoted in Salinas and Cartwright, "Welcoming and Opening Remarks," in Douglas W. Shaw, Aldon, and LoSapio, *Desired Future Conditions*, 2.
18. Aldo Leopold, *A Sand County Almanac* (New York: Ballantine Books, 1991), xvii.
19. Aldo Leopold, "Grass, Brush, Timber, and Fire in Southern Arizona," *Journal of Forestry* 22 (1924); Julianne Lutz Newton, *Aldo Leopold's Odyssey: Rediscovering the Author of a Sand County Almanac.* (Washington, D.C.: Island Press, 2006), 66–69.
20. Newton, *Aldo Leopold's Odyssey*, 73–74.
21. Quoted in Sid Goodloe, "Integrated Resource Management on Public and Private Rangelands," in Douglas W. Shaw, Aldon, and LoSapio, *Desired Future Conditions*, 135.
22. Ibid.
23. Patoski, "War on Cedar."
24. Newton, *Aldo Leopold's Odyssey*, 13–15.
25. Howard Wilshire, *The American West at Risk: Science, Myths, and the Politics of Land Abuse and Recovery* (Oxford: Oxford University Press, 2008), 135.

26. Aldo Leopold, like many others, viewed the earth as one system. "[It] is not impossible to regard the earth's parts—soil, mountains, rivers, atmosphere, etc.—as organs, or parts of organs, of a coordinated whole," he wrote. "Some Fundamentals of Conservation in the Southwest," *Environmental Ethics* 1 (Summer 1979): 139.
27. Romme et al., "Historical and Modern Disturbance," 217.
28. Ibid., 215, 217.
29. Also known as microbiotic, cryptogamic, microphytic, or cryptobiotic crusts.
30. Quoted in Michelle Nijhuis, "Getting under the Desert's Skin: Biologist Jayne Belnap," *High Country News*, January 19, 2004.
31. Jayne Belnap, comments, September 13, 2013, Canyonlands National Park. Others have also argued that carbon dioxide benefits junipers more than it benefits grasses. See Rick Miller et al., "Western Juniper Woodlands."
32. Harley G. Shaw, *Wood Plenty, Grass Good*, 4.
33. Richard F. Miller, Robin J. Tausch, E. Durant McArthur, Dustin D. Johnson, and Stewart C. Sanderson, *Age Structure and Expansion of Piñon-Juniper Woodlands: A Regional Perspective in the Intermountain West*, Research Paper Report RMRS-RP-69 (U.S. Department of Agriculture, Forest Service, Rocky Mountain Research Station, 2008).

Chapter 9: Relationships

1. Salinas and Cartwright, "Welcoming and Opening Remarks," in Douglas W. Shaw, Aldon, and LoSapio, *Desired Future Conditions*, 3.
2. Nelson, *Juniper and Black Pine*, 38–39.
3. C. S. Lewis, *The Magician's Nephew* (New York: Harper Collins, 2002), 76.
4. Mark Brunson, interview with author, November 8, 2014.
5. Gwendolyn L. Waring, *A Natural History of the Intermountain West* (Salt Lake City: University of Utah Press, 2011), 99–100.
6. Allison Jones, Jim Catlin, and Emanuel Vasquez, *Mechanical Treatment of Pinyon-Juniper and Sagebrush Systems in the Intermountain West: A Review of the Literature* (Wild Utah Project report, March 2013), 3, accessed April 26, 2017, https://static1.squarespace.com/static/57c5f6aa579fb31d71581457/t/58be552fe58c6278e8ad3d2a/1488868657789/MechTrt_LitReview.pdf.
7. Frischknecht, "Native Faunal Relationships," 64–65.
8. Don D. Dwyer, "Response of Livestock Forage to Manipulation of the Pinyon-Juniper Ecosystem," in *Pinyon-Juniper Ecosystem*, 98.
9. Bruce K. Koyiyumptewa, "Spiritual Values of the Piñon-Juniper Woodland: A Hopi Perspective," in Aldon and Shaw, *Managing Piñon-Juniper Ecosystems*, 19.
10. Kareen Shaheen, "Protecting Lebanon's Other Tree: The Juniper," *Daily Star* (Lebanon), August 24, 2013, accessed October 1, 2014, http://www.dailystar.

com.lb/Culture/Lifestyle/2013/Aug-24/228530-protecting-lebanons-other-tree-the-juniper.ashx#axzz3EvPKuOev.

11. Dwyer, "Response of Livestock Forage," 97.
12. Lanner and Frazier, "Historical Stability," 361.
13. Neil E. West, "Successional Patterns and Productivity Potentials of Pinyon-Juniper Ecosystems," in *Developing Strategies for Rangeland Management*, National Research Council/National Academy of Sciences Report (Boulder, CO: Westview Press, 1984), 1316; Fee Busby, professor of wildland resources, Utah State University, telephone interview with author, September 17, 2014.
14. Richard S. Aro, "Pinyon-Juniper Woodland Manipulation with Mechanical Methods," in *Pinyon-Juniper Ecosystem*, 67–68.
15. Busby, interview.
16. Comment to Richard Stevens, Bruce C. Giunta, and A. Perry Plummer, "Some Aspects in the Biological Control of Juniper and Pinyon," in *Pinyon-Juniper Ecosystem*, 90.
17. Many citizens consider PJ to be "one of our most impressive vegetation types aesthetically." See Dwyer, "Response of Livestock Forage," 98.
18. Quoted in Aro, "Pinyon-Juniper Woodland Manipulation," 67.
19. Jones, Catlin, and Vasquez, *Mechanical Treatment of Pinyon-Juniper*, 2.
20. Frischknecht, "Native Faunal Relationships," 55–60; Waring, *Natural History of the Intermountain West*, 105–9.
21. Warren P. Clary, "Present and Future Multiple Use Demands on the Pinyon-Juniper Type," in *Pinyon-Juniper Ecosystem*, 21.
22. Clary, "Present and Future Multiple Use," 22.
23. West, "Successional Patterns," 1317.
24. Aro, "Pinyon-Juniper Woodland Manipulation," 69.
25. West, "Successional Patterns," 1317.
26. "Controversial Technique Improves Land," *Bulletin* (Bend, OR), August 29, 1979, Wagner Mall insert, 2, accessed December 8, 2014, http://news.google.com/newspapers?nid=1243&dat=19790828&id=pNlYAAAAIBAJ&sjid=__YDAAAAIBAJ&pg=6625,3754043.
27. Deboodt et al., "Monitoring Hydrological Changes," 227.
28. Gerald F. Gifford, "Impacts of Pinyon-Juniper Manipulation on Watershed Values," in *Pinyon-Juniper Ecosystem*, 141. See also Hurst, "Management Strategies," in *Pinyon-Juniper Ecosystem*, 193. Hurst, regional forester for the Southwest Region, said, "For a long time it was my opinion and of others that we could improve water yield by removing p-js, but that theory has been disproved."
29. Brunson, interview.

30. Joseph Bauman, "Groups Raise Cain over Plan to Raze Kane Trees," *Deseret News*, June 11, 1991.
31. Lee E. Hughes, "Range Management and Image," *Rangelands* 14 (August 1992).
32. "2 Federal Agencies Drift Apart over Chaining Trees," *Deseret News*, September 6–7, 1994.
33. Brunson, interview.
34. Doug Page, interview with author, November 5, 2014.
35. Allison Jones, Wild Utah, interview with author, September 19, 2014.
36. Busby, interview.
37. Neal Clark, interview with author, September 10, 2014.
38. Keith Olive, interview with author, July 16, 2010.
39. Evan I. DeBloois, Dee F. Green, and Henry G. Wylie, "A Test of the Impact of Pinyon-Juniper Chaining on Archaeological Sites," in *Pinyon-Juniper Ecosystem*, 161.
40. Lori Hunsaker, personal communication, September 8, 2014.
41. Busby, interview. For an account of wildfires in 1996, see "Western Wildfires Chase Away Tourists, Residents," *CNN*, August 19, 1996, accessed September 23, 2014, http://www.cnn.com/US/9608/19/wildfires/. In one weekend in August 1996, hundreds of thousands of acres were burning in five states, including a dozen fires in Colorado alone. The next year was another bad year. Among the 620,730 acres that burned in Utah in 1997, the massive Milford Flat Fire blackened 363,052 acres.
42. Lori Hunsaker, personal communication, September 8, 2014.
43. Tueller and Clark, "Autecology of Pinyon-Juniper Species," 36. See also Donald A. Jameson, "Degradation and Accumulation of Inhibitory Substances from *Juniperus osteosperma* (Torr.) Little," in *Biochemical Interactions among Plants* (Washington, D.C.: National Academy of Sciences, 1971), 121–27.
44. Read, "I, Pencil."
45. Penny Frazier, comment on "Pinyon-Juniper Chaining Project Will Benefit Wildlife, Watershed," *Deseret News*, November 8, 2007; comment published online November 14, 2007, accessed September 24, 2014, http://www.deseretnews.com/article/695225345/Pinyon-juniper-chaining-project-will-benefit-wildlife-watershed.html?pg=all.
46. Vickie Tyler, interview with author, May 24, 2010.
47. Vern Huser and Paul Rokich, "Environmental Concerns of Pinyon-Juniper Management," in *Pinyon-Juniper Ecosystem*, 179. Comment, 184.

Chapter 10: Restoration

1. Vickie Tyler, interview.

2. Tiffany Bartz, interview with author, May 24, 2010; "Upper Kanab Creek Watershed Vegetation Management Project, Environmental Assessment #UT-040-09-03," accessed November 2, 2016, http://action.suwa.org/site/DocServer/FactSheet_UpperKanbCreekVegTreatment.pdf?docID=10061.
3. Quoted in Turner, *Earth's Blanket*, 73.
4. Some of the research is explained in Peter Wohlleben, *The Hidden Life of Trees: What They Feel, How They Communicate; Discoveries from a Secret World* (Vancouver: Greystone Books, 2016).
5. Miranda D. Redmond, Neil S. Cobb, Mark E. Miller, and Nichole N. Barger, "Long-Term Effects of Chaining Treatments on Vegetation Structure in Piñon-Juniper Woodlands of the Colorado Plateau," *Forest Ecology and Management* 305 (October 2013): 120–28; Nathan A. Bristow, Peter J. Weisberg, and Robin J. Tausch, "A 40-Year Record of Tree Establishment following Chaining and Prescribed Fire Treatments in Singleleaf Pinyon (*Pinus monophylla*) and Utah Juniper (*Juniperus osteosperma*) Woodlands," *Rangeland Ecology and Management* 67, no. 4 (2014): 389–96.
6. *Oxford English Dictionary* (Oxford: Oxford University Press, 2014), accessed at Salt Lake Public Library website, December 11, 2014, http://www.oed.com.ezproxy.slcpl.org/view/Entry/163986?redirectedFrom=restoration#eid.
7. Belnap, interview.
8. Mark Brunson offered this analogy in an interview.
9. U.S. Department of the Interior Bureau of Land Management, Environmental Assessment UT-Y020-2011-0047-EA, October 2012, Beef Basin/Dark Canyon Plateau Sagebrush Restoration, 31. Copy in possession of author.
10. Johnsen, "One-Seed Juniper Invasion," 204–5.
11. Ronald Lanner, e-mail message to author, February 2, 2012.
12. Clark, interview.
13. Patoski, "War on Cedar."
14. Catherine S. Fowler, "'We Live by Them': Native Knowledge of Biodiversity in the Great Basin of Western North America," in *Biodiversity and Native America*, ed. Paul E. Minnis and Wayne J. Elisens (Norman: University of Oklahoma Press, 2001), 115–18; M. Kat Anderson, *Tending the Wild: Native American Knowledge and the Management of California's Natural Resources* (Berkeley: University of California Press, 2005), 316.
15. Anderson, *Tending the Wild*, 284. The speaker is probably referring to Crowley Lake, a reservoir in Mono County, California, and to the landscape before its creation in 1941.
16. Trudeau, *Environmental History*, 117–24.
17. Comment to Stevens, Giunta, and Plummer, "Aspects in the Biological Control," in *Pinyon-Juniper Ecosystem*, 103.

18. Johnson, "Pinyon-Juniper Forests," 121.
19. Baker and Shinneman, "Fire and Restoration," 18.
20. See Baker and Shinneman, "Fire and Restoration," 1–21.
21. Romme, Floyd-Hanna, and Hanna, "Ancient Piñon-Juniper Forests," 335–36.
22. Romme et al., "Historical and Modern Disturbance," 204.
23. Turner, *Earth's Blanket*, 144–45.
24. Sharon Baruch-Mordo, Jeffrey S. Evans, John P. Severson, David E. Naugle, Jeremy D. Maestas, Joseph M. Kiesecker, Michael J. Falkowski, Christian A. Hagen, and Kerry P. Reese, "Saving Sage-Grouse from the Trees: A Proactive Solution to Reducing a Key Threat to a Candidate Species," *Biological Conservation* 167 (2013): 233–41; Jones, interview. Sage grouse avoid conifers during nesting, brood rearing, and wintering, and even a 4 percent tree cover impacts these birds; they don't like being around places where raptors can perch.
25. Belnap, interview.
26. Romme, Floyd-Hanna, and Hanna, "Ancient Piñon-Juniper Forests," 335–36. These ideas are also noted in Belnap, interview; Busby, interview; Jones, Catlin, and Vasquez, *Mechanical Treatment of Pinyon-Juniper*, 5. For example, an existing study from a Shay Mesa treatment project shows an increase in "vegetative cover" but goes on to state that this cover consists of nondesirable, nonnative species. See *Shay Mesa: Trend Study No. 14-11-09*, accessed October 14, 2014, https://wildlife.utah.gov/range/pdf/wmu14/wmu14-11.pdf.
27. Jones, Catlin, and Vasquez, *Mechanical Treatment of Pinyon-Juniper*, 5.
28. Hurst, "Management Strategies," 193.
29. In the 1970s treatments were such that forester William Hurst pointed out that a treatment might increase forage for two or three years, and then "it falls off to nothing." He noted that Bruce King, a cattleman and former governor of New Mexico, wanted to trade some of his treated land for the Forest Service's untreated land, because the untreated lands had more forage. Of course, the management of grazing posttreatment also contributes to the long-term condition of the range. Hurst, "Management Strategies."
30. Belnap, interview.
31. Jones, interview.
32. Earl C. Hindley, James Bowns, Edward Scherick, Paul Curtis, and Jimmie Forrest, *A Photographic History of Vegetation and Stream Channel Changes in San Juan County, Utah* (San Juan County, Charles Redd Foundation, Utah State University Extension, 2000), 9ff. Carefully managed grazing has been used in combination with other strategies to improve and restore land—for instance, at the North Rim ranches managed by the Grand Canyon Trust—but an exploration of this strategy is beyond the scope of this work.

33. Terence P. Yorks, Neil E. West, and Kathleen M. Capels, "Changes in Pinyon-Juniper Woodlands in Western Utah's Pine Valley between 1933–1989," *Journal of Range Management* 47 (September 1994): 359–64.
34. Clark, interview.
35. Baker and Shinneman, "Fire and Restoration," 18.
36. James A. Harris, Richard J. Hobbs, Eric S. Higgs, and James Aronson, "Ecological Restoration and Global Climate Change," in *Restoration Ecology* 14 (June 2006): 171.
37. Jayne Belnap, comments, Canyonlands.
38. Belnap, interview.
39. Harris et al., "Ecological Restoration," 171.

CHAPTER 11: KIN

1. Carl Jung, "Letter to Dr. S.," October 8, 1947, in *C. G. Jung Letters*, vol. 1, *1906–1950*, ed. Gerhard Adler (Princeton, NJ: Princeton University Press), 479.
2. Mark Brunson, "Unwanted No More: Land Use, Ecosystem Services, and Opportunities for Resilience in Human-Influenced Shrublands," *Rangelands* 36 (April 2014): 6–8.
3. Michael Shermer, "Life Has Never Been So Good for Our Species," *Los Angeles Times*, April 30, 2010.
4. Ken Cole, "War on Trees: Harry Reid, Ag Extension Agents, and Chinese Biomass Companies Promote Liquidation of Old Growth Forests in Nevada," *Wildlife News*, January 18, 2011; Katie Fite, Western Watersheds Project, e-mail message to author, November 10, 2014.
5. John Muir, *Our National Parks* (Boston: Houghton Mifflin, 1901), 1.
6. Florence Williams, "Take Two Hours of Pine Forest and Call Me in the Morning," *Outside*, December 2012. See also M. Amos Clifford, *A Little Handbook of Shinrin-Yoku* (Association of Nature and Forest Therapy Guides and Programs, 2013).
7. Carlo Rovelli, "All Reality Is Interaction," interview with Krista Tippett, *On Being*, accessed May 5, 2017, https://onbeing.org/programs/carlo-rovelli-all-reality-is-interaction/. If I understand it right, this idea is part of what the Hindu image of Indra's net, thousands of years old, is expressing.

Selected Bibliography

Abbey, Edward. *Desert Solitaire.* New York: Ballantine Books, 1971.

Adams, Karen R. "Subsistence and Plant Use during the Chacoan and Second Occupations at Salmon Ruin." In *Chaco's Northern Prodigies: Salmon, Aztec and the Ascendance of the Middle San Juan Region after AD 1100*, edited by Paul F. Reed, 65–85. Salt Lake City: University of Utah Press, 2008.

Adams, Robert P. *Junipers of the World: The Genus* Juniperus. 4th ed. Bloomington, IN: Trafford, 2014.

———. "Yields and Seasonal Variation of Phytochemicals from *Juniperus* Species of the United States." *Biomass* 12, no. 2 (1987): 129–39.

Aldon, Earl F., and Douglas W. Shaw, tech. coordinators. *Managing Piñon-Juniper Ecosystems for Sustainability and Social Needs.* General Technical Report RM-236. Fort Collins, CO: U.S. Department of Agriculture, Forest Service, Rocky Mountain Forest and Range Experiment Station, 1993.

Anderson, M. Kat. *Tending the Wild: Native American Knowledge and the Management of California's Natural Resources.* Berkeley: University of California Press, 2005.

Arendt, Paul A., and William L. Baker. "Northern Colorado Plateau Piñon-Juniper Woodland Decline over the Past Century." *Ecosphere* 4 (August 2013). doi: 10.1890/ES13-00081.1.

Baker, William L., and Douglas J. Shinneman. "Fire and Restoration of Piñon-Juniper Woodlands in the Western United States: A Review." *Forest Ecology and Management* 189 (2004): 1–21.

Barton, John D. *A History of Duchesne County.* Salt Lake City: Utah State Historical Society and Duchesne County Commission, 1998.

Bate, Kerry William. "Iron City, Mormon Mining Town." *Utah Historical Quarterly* 50 (Winter 1983): 47–58.

Bauer, John M., and Peter J. Weisberg. "Fire History of a Central Nevada Pinyon-Juniper Woodland." *Canadian Journal of Forest Research* 39 (August 2009): 1589–99. doi: 10.1139/X09-078.

Born, J. David, Ronald P. Tymcio, and Osborne E. Casey. *Nevada's Forest Resources.* Resource Bulletin INT-76. Fort Collins, CO: U.S. Department of Agriculture, Forest Service, Intermountain Research Station, July 1992.

Bradley, Martha Sonntag. *A History of Beaver County.* Salt Lake City: Utah State Historical Society and Beaver County Commission, 1999.

Bristow, Nathan A., Peter J. Weisberg, and Robin J. Tausch. "A 40-Year Record of Tree Establishment following Chaining and Prescribed Fire Treatments in Singleleaf Pinyon (*Pinus monophylla*) and Utah Juniper (*Juniperus osteosperma*) Woodlands." *Rangeland Ecology and Management* 67, no. 4 (2014): 389–96.

Brunson, Mark. "Unwanted No More: Land Use, Ecosystem Services, and Opportunities for Resilience in Human-Influenced Shrublands." *Rangelands* 36 (April 2014): 5–11.

Bryant, Edwin. *What I Saw in California, Being the Journal of a Tour, in the Years 1846, 1847.* New York: D. Appleton, 1849.

Burroughs, John. *The Writings of John Burroughs.* Vol. 17. Boston: Houghton Mifflin, 1913.

Clifford, M. Amos. *A Little Handbook of Shinrin-Yoku.* Association of Nature and Forest Therapy Guides and Programs, 2013.

Coats, Larry L., Kenneth L. Cole, and Jim I. Mead. "50,000 Years of Vegetation and Climate History on the Colorado Plateau, Utah and Arizona, USA." *Quaternary Research* 70, no. 2 (2008): 322–38.

Cook, Theodore Andrea. *The Curves of Life.* London: Constable, 1914.

Coues, Elliott, ed. and trans. *On the Trail of a Spanish Pioneer: The Diary and Itinerary of Francisco Garcés.* Vol. 2. New York: Francis P. Harper, 1900.

Culpeper, Nicholas. *Culpeper's Complete Herbal: With Nearly Four Hundred Medicines, Made from English Herbs, Physically Applied to the Cure of All Disorders Incident to Man; with Rules for Compounding Them; Also, Directions for Making Syrups, Ointments, &C.* London: Milner and Sowerby, 1852.

DeBlander, Larry T., John D. Shaw, Chris Witt, Jim Menlove, Michael T. Thompson, Todd A. Morgan, R. Justin DeRose, and Michael C. Amacher. *Utah's Forest Resources, 2000–2005.* Resource Bulletin RMRS-RB-10. Fort Collins, CO: U.S. Department of Agriculture, Forest Service, Rocky Mountain Research Station, 2010.

Deboodt, T. L., M. P. Fisher, J. C. Buckhouse, and J. Swanson. "Monitoring Hydrological Changes Related to Western Juniper Removal: A Paired Watershed Approach." In *Planning for an Uncertain Future—Monitoring, Integration, and Adaptation*, edited by Richard M. T. Webb and Darius J. Semmens, 227–32. Proceedings of the Third Interagency Conference on Research in the Watersheds, September 8–11, 2008, Estes Park, CO. Scientific Investigations Report 2009-5049. Reston, VA: U.S. Geological Survey, 2009.

Dowling, Alfred E. P. Raymond. *The Flora of the Sacred Nativity.* London: Kegan Paul, Trench, Trubner, 1900.

Elmore, Francis H. *Ethnobotany of the Navajo.* Monograph No. 8. Santa Fe: University of New Mexico and School of American Research, July 1944.

Erdman, James A. "Pinyon-Juniper Succession after Natural Fires on Residual Soils of Mesa Verde, Colorado." Pamphlet 13112, Utah History Research Center. Provo, UT: Brigham Young University, 1970.

Fowler, Catherine S. "'We Live by Them': Native Knowledge of Biodiversity in the Great Basin of Western North America." In *Biodiversity and Native America*, edited by Paul E. Minnis and Wayne J. Elisens, 99–132. Norman: University of Oklahoma Press, 2001.

Geary, Edward A. *A History of Emery County.* Salt Lake City: Utah State Historical Society and Emery County Commission, 1996.

Goeking, Sara A., John D. Shaw, Chris Witt, Michael T. Thompson, Charles E. Werstak Jr., Michael C. Amacher, Mary Stuever, et al. *New Mexico's Forest Resources, 2008–2012.* Resource Bulletin RMRS-RB-18. Fort Collins, CO: U.S. Department of Agriculture, Forest Service, Rocky Mountain Research Station, August 2014.

Graves, Henry S., Frank J. Phillips, and Walter Mulford. *Utah Juniper in Central Arizona.* Circular 197. U.S. Department of Agriculture, Forest Service. Washington, D.C.: Government Printing Office, 1912.

Grimm, Jacob. *Teutonic Mythology*. Vol. 2. Translated by James Steven Stallybrass. London: George Bell and Sons, 1883.

Hampe, Arndt, and Rémy J. Petit. "Cryptic Forest Refugia on the 'Roof of the World.'" *New Phytologist* 185 (December 2009). doi:10.1111/j.1469-8137.2009.03112.x.

Harris, James A., Richard J. Hobbs, Eric S. Higgs, and James Aronson. "Ecological Restoration and Global Climate Change." *Restoration Ecology* 14 (June 2006). doi:10.1111/j.1526-100X.2006.00136.x.

Hindley, Earl C., James Bowns, Edward Scherick, Paul Curtis, and Jimmie Forrest. *A Photographic History of Vegetation and Stream Channel Changes in San Juan County, Utah.* Supported by San Juan County, Charles Redd Foundation, and Utah State University Extension, 2000.

Jackson, Steve, and Julio Betancourt. *Late Holocene Expansion of Utah Juniper in Wyoming: A Modeling System for Studying Ecology of Natural Invasions.* NSF-DEB-9815500. Final Report, Collaborative Research, National Science Foundation, n.d. https://wwwpaztcn.wr.usgs.gov/wyoming/NSF_report.pdf.

Jameson, Donald A. "Degradation and Accumulation of Inhibitory Substances from *Juniperus osteosperma* (Torr.) Little." In *Biochemical Interactions among Plants*, 121–27. Washington, D.C.: National Academy of Sciences, 1971.

Janetski, Joel C., Karen D. Lupo, John M. McCullough, and Shannon A. Novak. "The Mosida Site: A Middle Archaic Burial from the Eastern Great Basin." *Journal of California and Great Basin Anthropology* 14, no. 2 (1992): 180–200.

Johnsen, Thomas N., Jr. "One-Seed Juniper Invasion of Northern Arizona Grasslands." *Ecological Monographs* 32 (Summer 1962): 187–207.

Ko, Dongwook W., Ashley D. Sparrow, and Peter J. Weisberg. "Land-Use Legacy of Historical Tree Harvesting for Charcoal Production in a Semi-arid Woodland." *Forest Ecology and Management* 261 (April 2011): 1283–92.

Kozlowski, T. T. *Growth and Development of Trees*. Vol. 2, *Cambial Growth, Root Growth, and Reproductive Growth*. New York: Academic Press, 1971.

Kubler, Hans. "Function of Spiral Grain in Trees." *Trees* 5, no. 3 (1991): 125–35.

Lanner, Ronald M. *The Piñon Pine: A Natural and Cultural History*. Reno: University of Nevada Press, 1981.

———. *Trees of the Great Basin: A Natural History*. Reno: University of Nevada Press, 1984.

Lanner, Ronald M., and Penny Frazier. "The Historical Stability of Nevada's Pinyon-Juniper Forest." *Phytologia* 93 (December 2011): 360–87.

Leopold, Aldo. "Grass, Brush, Timber, and Fire in Southern Arizona." *Journal of Forestry* 22 (1924): 1–10.

Linton, M. J., J. S. Sperry, and D. G. Williams. "Limits to Water Transport in *Juniperus osteosperma* and *Pinus edulis*: Implications for Drought Tolerance and Regulation of Transpiration." *Functional Ecology* 12, no. 6 (1998): 906–11.

Lyford, Mark E., Stephen T. Jackson, Julio L. Betancourt, and Stephen T. Gray. "Influence of Landscape Structure and Climate Variability on a Late Holocene Plant Migration." *Ecological Monographs* 73, no. 4 (2003): 567–83.

Mayes, Vernon O., and Barbara Bayless Lacy. *Nanise': A Navajo Herbal; One Hundred Plants from the Navajo*. Tsaile, AZ: Navajo Community College Press, 1989.

Miller, David E. *Hole in the Rock: An Epic Colonization of the Great American West*. Salt Lake City: University of Utah Press, 1966.

Miller, Richard F., Robin J. Tausch, E. Durant McArthur, Dustin D. Johnson, and Stewart C. Sanderson. *Age Structure and Expansion of Piñon-Juniper Woodlands: A Regional Perspective in the Intermountain West*. Research Paper Report RMRS-RP-69. Fort Collins, CO: U.S. Department of Agriculture, Forest Service, Rocky Mountain Research Station, 2008.

Moerman, Daniel. *Native American Ethnobotany*. Portland, OR: Timber Press, 1998.

Morgan, Dale. *Jedediah Smith and the Opening of the West*. Lincoln, NE: Bison Books, 1964.

Morris, Patrick C. "Bears, Juniper Trees, and Deer, the Metaphors of Domestic Life, an Analysis of a Yavapai Variant of the Bear Maiden Story." *Journal of Anthropological Research* 32 (Autumn 1976): 246–54.

Muir, John. *Our National Parks.* Boston: Houghton Mifflin, 1901.

Murphy, Miriam B. *A History of Wayne County.* Salt Lake City: Utah State Historical Society and Wayne County Commission, 1999.

Native American Ethnobotany: A Database of Foods, Drugs, Dyes and Fibers of Native American Peoples, Derived from Plants. University of Michigan–Dearborn. http://naeb.brit.org/.

Nelson, Lillian Barrus. *Juniper and Black Pine: A History of Two Southern Idaho Communities, 1870s to 1995.* Salt Lake City: Publishers Press, 1996.

Newell, Linda King, and Vivian Linford Talbot. *A History of Garfield County.* Salt Lake City: Utah State Historical Society and Garfield County Commission, 1998.

Newton, Julianne Lutz. *Aldo Leopold's Odyssey: Rediscovering the Author of a Sand County Almanac.* Washington, D.C.: Island Press, 2006.

Nowak, Cheryl L., Robert S. Nowak, Robin J. Tausch, and Peter E. Wigand. "Tree and Shrub Dynamics in Northwestern Great Basin Woodland and Shrub Steppe during the Late-Pleistocene and Holocene." *American Journal of Botany* 88 (March 1994): 265–77.

O'Brien, Renee A. *Arizona's Forest Resources, 1999.* Resource Bulletin RMRS-RB-2. Fort Collins, CO: U.S. Department of Agriculture, Forest Service, Rocky Mountain Research Station, 2002.

Peattie, Donald Culross. *A Natural History of Western Trees.* Boston: Houghton Mifflin, 1953.

The Pinyon-Juniper Ecosystem: A Symposium, May 1975. Logan: Utah State University, College of Natural Resources, Utah Agricultural Experiment Station, 1975.

Redmond, Miranda D., Neil S. Cobb, Mark E. Miller, and Nichole N. Barger. "Long-Term Effects of Chaining Treatments on Vegetation Structure in Piñon-Juniper Woodlands of the Colorado Plateau." *Forest Ecology and Management* 305 (October 2013): 120–28.

Reed, Paul F., ed. *Chaco's Northern Prodigies: Salmon, Aztec and the Ascendance of the Middle San Juan Region after AD 1100.* Salt Lake City: University of Utah Press, 2008.

Reiner, Alicia L. "Fuel Load and Understory Community Changes Associated with Varying Elevation and Pinyon-Juniper Dominance." Master's thesis, University of Nevada–Reno, 2004.

Romme, William H., Craig D. Allen, John D. Bailey, William L. Baker, Brandon T. Bestelmeyer, Peter M. Brown, Karen S. Eisenhart, et al. "Historical and Modern Disturbance Regimes, Stand Structures, and Landscape Dynamics in

Piñon-Juniper Vegetation of the Western United States." *Rangeland Ecology and Management* 62 (May 2009): 203–22.

Romme, William H., Lisa Floyd-Hanna, and David D. Hanna. "Ancient Piñon-Juniper Forests of Mesa Verde and the West: A Cautionary Note for Forest Restoration Programs" In *Fire, Fuel Treatments, and Ecological Restoration: Conference Proceedings, April 16–18, 2002*, edited by Philip N. Omi and Linda A. Joyce, 335–50. Proceedings RMRS-P-29. Fort Collins, CO: U.S. Department of Agriculture, Forest Service, Rocky Mountain Research Station, 2003.

Schulgasser, K., and A. Witztum. "The Mechanism of Spiral Grain Formation in Trees." *Wood Science Technology* 41, no. 2 (2007): 133–56.

Shaw, Douglas W., Earl F. Aldon, and Carol LoSapio, tech. coordinators. *Desired Future Conditions for Piñon-Juniper Ecosystems.* General Technical Report RM-258. Fort Collins, CO: U.S. Department of Agriculture, Forest Service, Rocky Mountain Forest and Range Experiment Station, 1995.

Shaw, Harley G. *Wood Plenty, Grass Good, Water None: Vegetation Changes in Arizona's Upper Verde River Watershed from 1850 to 1997.* General Technical Report RMRS-GTR-177. Fort Collins, CO: U.S. Department of Agriculture, Forest Service, Rocky Mountain Research Center, 2006.

Shirts, Morris A. "The Iron Mission." In *Utah History Encyclopedia,* edited by Allan Kent Powell. Salt Lake City: University of Utah Press, 1994.

Simpson, Georgiana Kennedy. *Navajo Ceremonial Baskets: Sacred Symbols, Sacred Space.* Summertown, TN: Native Voices, 2003.

Skatter, Sondre, and Bohumil Kucera. "The Cause of the Prevalent Directions of the Spiral Grain Patterns in Conifers." *Trees* 12, no. 5 (1998): 265–73.

Small, Ernest. *North American Cornucopia: Top 100 Indigenous Food Plants.* Boca Raton, FL: CRC Press, 2013.

Smart, William B. *Old Utah Trails.* Salt Lake City: Utah Geographic Series, 1988.

Smart, William B., and Donna T. Smart. *Over the Rim: The Parley P. Pratt Exploring Expedition to Southern Utah, 1849–1850.* Logan: Utah State University Press, 1999.

Stansbury, Howard. *An Expedition to the Valley of the Great Salt Lake of Utah.* Philadelphia: Lippincott, Grambo, 1855.

Strachan, Scotty, Franco Biondi, Susan G. Lindström, Robert McQueen, and Peter E. Wigand. "Application of Dendrochronology to Historical Charcoal-Production Sites in the Great Basin, United States." *Historical Archaeology* 47, no. 4 (2013): 103–19.

Straka, Thomas J., and Robert H. Wynn. "History on the Road: Charcoal and Nevada's Early Mining History." *Forest History Today*, Fall 2008, 63–66.

Tellman, Barbara, and Diana Hadley. *Crossing Boundaries: An Environmental History of the Upper San Pedro River Watershed, Arizona and Sonora.* Tucson: Office of Ethnohistorical Research, Arizona State Museum, University of Arizona, 2006.

Toelken, Barre. "Seeing with a Native Eye: How Many Sheep Will It Hold?" In *Seeing with a Native Eye: Essays on Native American Religion*, edited by Walter Holden Capps, 9–24. San Francisco: Harper, 1976.

Turner, Nancy. *The Earth's Blanket: Traditional Teachings for Sustainable Living.* Seattle: University of Washington Press, 2005.

Waring, Gwendolyn L. *A Natural History of the Intermountain West.* Salt Lake City: University of Utah Press, 2011.

West, Neil E. "Successional Patterns and Productivity Potentials of Pinyon-Juniper Ecosystems." In *Developing Strategies for Rangeland Management*, 1301–32. National Research Council/National Academy of Sciences Report. Boulder, CO: Westview Press, 1984.

Wilcox, Bradford P., and Thomas L. Thurow. Preface to "Emerging Issues in Rangeland Ecohydrology." Special issue, *Hydrological Processes* 20, no. 15 (2006): 3155–57.

Williams, Gerald W., comp. "References on the American Indian Use of Fire in Ecosystems." Washington, D.C.: U.S. Department of Agriculture, Forest Service, 2005. https://www.nrcs.usda.gov/Internet/FSE_DOCUMENTS/nrcs144p2_051334.pdf.

Willson, Cynthia J., Paul S. Manos, and Robert B. Jackson. "Hydraulic Traits Are Influenced by Phylogenetic History in the Drought-Resistant, Invasive Genus *Juniperus* (Cupressaceae)." *American Journal of Botany* 95 (March 2008): 299–314.

Wilshire, Howard. *The American West at Risk: Science, Myths, and the Politics of Land Abuse and Recovery.* Oxford: Oxford University Press, 2008.

Windes, Thomas C., and Eileen Bacha. "Sighting along the Grain: Differential Wood Use at Salmon Ruin." In *Chaco's Northern Prodigies: Salmon, Aztec and the Ascendancy of the Middle San Juan Region after AD 1100*, edited by Paul F. Reed, 113–39. Salt Lake City: University of Utah Press, 2008.

Wohlleben, Peter. *The Hidden Life of Trees: What They Feel, How They Communicate; Discoveries from a Secret World.* Vancouver: Greystone Books, 2016.

Yorks, Terence P., Neil E. West, and Kathleen M. Capels. "Changes in Pinyon-Juniper Woodlands in Western Utah's Pine Valley between 1933–1989." *Journal of Range Management* 47 (September 1994): 359–64.

Young, Karl. "Wild Cows of the San Juan." *Utah Historical Quarterly* 32 (July 1964): 251–67.

Index

Page numbers in italics refer to images.